Burt Franklin: Research & Source Work Series #184
(American Classics in History & Social Science #27)

AMERICAN FACTORIES

AND THEIR

FEMALE OPERATIVES

AMERICAN FACTORIES

AND THEIR

FEMALE OPERATIVES;

WITH AN APPEAL ON BEHALF OF THE

BRITISH FACTORY POPULATION,

AND SUGGESTIONS FOR THE IMPROVEMENT
OF THEIR CONDITION.

BY THE REV. WILLIAM SCORESBY

Burt Franklin: Research & Source Work Series #184
(American Classics in History & Social Science #27)

BURT FRANKLIN
New York, N. Y.
1968

Published By
BURT FRANKLIN
235 East 44th St.
New York, N.Y. 10017

ORIGINALLY PUBLISHED
BOSTON: 1845
From the London Edition

Reprinted 1968

This work has been reprinted on
long-life paper.

Printed in U.S.A.

ADVERTISEMENT.

A visit by the Author to the United States and Canada, in the summer of 1844, afforded the materials for a series of Lectures on America, which were delivered, from time to time, before the members of the Bradford Church Institution. In the progress of the series, two Lectures were given, — on the evenings of the 3d and 17th of February last, — on the subject of the "American Factories and their Female Operatives." On one of these occasions, when there was an assemblage of about a thousand persons, a resolution, on the motion of a principal mill-owner, was unanimously passed, requesting the publication of the statements on a subject deemed of peculiar importance in a district, like that of Bradford, in which such a vast population is employed in its extensive mills and manufactories.

In compliance with this request, the following

pages, comprising the substance of these two Lectures, recast and considerably enlarged, have been printed, and are now submitted to the public. Truly grateful will the Author feel to a gracious and beneficent Providence, if the information herein embodied, and the practical suggestions drawn therefrom, should prove the occasion of calling public attention to the subject of the improvement of the condition of our female factory operatives, and might be blessed as the humble means of promoting, in any degree, this great and truly Christian enterprise.

Bradford, Yorkshire,
April 3, 1845.

CONTENTS.

CHAPTER I.

CHAPTER II.

CHAPTER III.

CHAPTER IV.

CHAPTER V.

AMERICAN FACTORIES

AND THEIR

FEMALE OPERATIVES.

CHAPTER I.

PROGRESS OF AMERICAN MANUFACTURES.

THERE are four principal sources from whence the wealth and prosperity of a country are derived, — from the natural and agricultural products of the country in its soil and mines — from manufactures — from fisheries, and from trade and commerce.

The first of these sources, made more effective by means of commercial interchange with other nations, is generally the earliest drawn upon in a newly settled country. The others become available more gradually, and especially manufactures, which can seldom become generally effective, or extend beyond mere domestic or handicraft productions — because of the expensive and complicated machinery required in them — until a colony has become in a very considerable degree matured.

Hence, for two hundred years after the planting of the original colonies in America, the chief source of national prosperity was in the soil; the produce of the soil being made the more valuable and enriching by commercial interchanges with England and other countries.

The Manufactures, which, in the early period of its

progress towards the condition of an independent nation, America engaged in, were such as were rather forced by the wants or necessities of her population, than by the freer influence of the principles of commercial enterprise.

The wars with this country — unhappy as, in our social relations and in respect to wisdom and humanity, these unnatural conflicts were—seemed to throw the Americans more upon their own resources, and thus to produce an influence, partly from necessity and partly from patriotic pride, stimulative of the native energies and mechanical talents.

Of late years, the start, which this and other causes had given to manufactures, has not only been promptly taken advantage of, but followed up with the ardor and enterprise which are so characteristic of the American nation.

Take an example in the case of cotton manufactures. In the year 1810, according to computations by a Committee of Congress, these manufactures in the United States consumed only 10,000 bales of the raw material; but, five years afterwards, the cotton worked up had increased to nine times that quantity. In 1835, the number of looms was estimated at 48,000; from 85 to 90 millions of pounds of cotton, or about 250,000 bales, were consumed, and the value of the produce was computed at 50,000,000 dollars.* In 1841-2, the number of cotton-mills in eight of the States of the Union — Maine, Massachusetts, Rhode Island, Connecticut, New York, Pennsylvania, Maryland, and Virginia — were estimated at 807:

* See an instructive article on the "United States," in the last edition of the Encyclopædia Britannica.

the amount of capital invested therein about forty millions of dollars: and the annual produce in cloth about 181,-400,000 yards. But in 1843 — such, apparently, has been the extraordinary advance — the factories of a single city, Lowell, were employing a capital of above ten millions of dollars, and were producing above seventy-four millions of yards of cotton cloth in the year! And so rapid has been the progress of manufactures in this one article of popular consumption, that we find, that, whereas the amount of cotton worked up in the whole of the Union was estimated in 1810 at only ten thousand bales, the manufactures of Lowell alone, in 1843, extended to nearly six times that quantity. And that, whereas in 1835, the whole of the cotton consumed throughout the entire country, in manufactures, was estimated at eighty-five to ninety millions of pounds of cotton,* in 1843, one-fourth of that quantity, or 22,880,000 pounds, was consumed in the mills of Lowell alone!

In the woollen manufactures of the United States, a progress in some degree correspondent, appears to have taken place. The State of New York alone numbered, in 1841, 258 woollen manufactories, yielding a produce of the value of about three millions of dollars.†

Iron manufactures have, in like manner, advanced in extent and importance — the value of their productions throughout the States, which in 1810 was estimated at 14,364,526 dollars, had, in 1840, advanced to four times that amount.

* For these figures I am indebted to secondary authorities, not having the opportunity of consulting the original documents. Whilst, therefore, I presume them to be correct, I would by no means answer for their precise accuracy.

† Davenport's American Gazetteer, p. 70.

Very recently, Professor Tucker of Virginia is stated to have given the following summary of the *annual* products of the United States: Agriculture, 654,387,597 dollars; commerce, 79,721,086; mining, 42,358,751; and manufactures, 239,836,224; giving a total of upwards of £200,000,000 sterling — of which nearly £50,000,000 belong to manufactures.

The rapid diminution of the American demands on the manufactures of this country agrees with and corroborates the foregoing results. Taking our *exports* to the United States, generally, we learn from returns laid before parliament in the session of 1843, that their annual value, instead of advancing with the growth and requirements of the population of that country, had rapidly diminished within a period of a few years. In 1832, the amount of our exports, in value, was £7,579,699. In 1842, when the value was the smallest of the previous ten years, the amount was only £3,528,107 — a reduction exceeding one-half. The greatest falling off was in cotton, woollen and silk manufactures.

Thus, the value of our exports to the United States in *cotton goods*, which in the year 1832 was £1,733,047, was only, in 1842, £482,276, or little more than one quarter of what it was ten years before. In woollen manufactures our exports, which in 1832 were of the value of £2,289,-883, had fallen, in 1842, to £892,335. And in silk, the proportional reduction in the value of our exports betwixt the years 1832 and 1842 was still greater — the value in 1832 being £351,278, and in 1842 only £81,243.*

Our *imports*, in the mean time, from the United States had greatly increased. The amount of cotton imported in

* American Almanac, for 1844, p. 302.

1832, was 237,506,758 lbs, and in 1842, 414,030,779 lbs. Of tobacco, in 1832, 20,748,317 lbs; in 1842, 38,618,012 lbs!

Thus we find the United States so changing its relations with Great Britain, as to receiver and producer, that we naturally infer that the time cannot be far distant when we shall have but little to contribute to her requirements, except the transfer of our surplus population towards the occupancy of her unreclaimed wilds, and to aid in the development of her vast and inexhaustible resources. Nor can we reasonably conceal from ourselves the probable results of her rapid advances in manufactures, in enabling her soon to rival ourselves, as, hitherto, the chief manufacturers for the world.

One grand difficulty, no doubt, which America has yet to contend with, is the greater cost, in labor, of her various fabrics; but, to set off against that, she has within herself the staple of most of her productions, with water-power, and fuel for raising steam-power, to an unlimited extent, besides cheapness of living for the hands required in her manufactories. With her, and for her prosperity, steam and machinery become paramount agencies. In her rich and exhaustless soil, no doubt, are the aliments of her growth and strength; but it is peculiarly her wisdom, too, to multiply, as far as enterprise reasonably may, the physical power in her existing deficiency of manual labor, by the aid of mechanical forces and still more extended machinery.

Instead of attempting to give you an account of the various places in the United States, which, in recent years, have grown up into manufacturing towns, or to describe the manner of conducting their various manufacturing establishments — which, for the space I have allotted to myself, would be an unprofitable, if not a hopeless task — I select the particular case of LOWELL, as formerly announced, which is decidedly the most important manufacturing town in the United States, except perhaps Pittsburg, for deliberate, and, I hope, useful consideration.

In selecting this particular case, however, as an example, I by no means adduce it, you will please to observe, as a *medium* or *average* representation of the American factory system in general. Of the condition of the factory operatives in other manufacturing towns in America, I know very little; but my impression is, that nowhere do they equal, in respect of their moral, social, and intellectual condition, what we find existing at Lowell.

But, with a practical and philanthropic object in view, I have selected this particular case, because, for a specimen on a large scale, I believe it to be the best existing pattern and example.

And this is my object — at least my most earnest desire and anxious prayerful hope — that, by describing to you the admirable arrangements of the factories in that city, and the remarkable superiority of character, acquirements and general conduct of the factory-girls of Lowell, our own factory masters and workers, and especially our female operatives, might be stimulated to take a lesson from them for the improvement of the manufacturing establishments of this town and neighborhood.

That you may have a clear, and as far as time will admit, a complete notion of the subject, I shall first give you some account of the city of Lowell, with its manufactories, &c.; next of the female operatives, or factory-girls employed there; then I shall inquire into the sources of their unusual moral and intellectual condition; and lastly, I shall offer some suggestions for the improvement of the condition of our factory operatives, especially of that of females.

These various particulars I shall endeavor to make so plain, that the youngest amongst you may be able, for the most part, to understand me.

CHAPTER II.

OF THE CITY OF LOWELL, WITH ITS MANUFACTURES, &c.

SECT. 1. In respect to the city of Lowell itself, it will be proper that I should describe to you, at the commencement, *Where it is situated.*

To enable the younger members of the Church Institution, and of my present audience generally, to obtain a correct idea of the position of Lowell, suppose we take an imaginary voyage thither, starting from Liverpool, in the royal-mail steam-packet.

From Liverpool (whence a steam-packet sails on the 4th and 18th of every month, two or three winter months excepted, when there is only a monthly packet) we proceed through the Irish Channel, close past the south part of Ireland, called Cape Clear, to which the distance will be about 350 English miles, and may take about 34 hours. Leaving Ireland on our right hand (the sailors call it the starboard hand) we go towards the west, a little southerly, for about 1700 nautical,* or 1969 English statute miles, which, in still weather, would take eight or nine days, and then come to Cape Race, the south-easternmost part of Newfoundland. Still going westerly, but inclining a little more to the south, we come in about two days or a little more, and after making an additional distance of 525 English miles, abreast of Halifax, in the British province of Nova Scotia. From this port, where mails and passen-

* This is calculated, not very accurately, on the principle of great-circle sailing; the distance, on the ordinary method of sailing, being longer, viz. about 1740 miles, and the course about W. 9° S.

gers are landed, we proceed in the same direction, very nearly, for, perhaps, 34 hours, a distance of about 370 miles, which brings us into Boston Bay, and an hour more into Boston harbor, where we land and proceed to the city.

We have thus completed our imaginary steam-voyage, extending to a distance of about 3225 English or statute miles from Liverpool, in a period of about thirteen days, an enterprise which, sailing from Plymouth, cost the pilgrim fathers, who were the first to colonize this very region, a dangerous and sickening voyage of about two months, or reckoning till their disembarkation, three months and a half.

Boston is the chief city of that part of the United States which is called New England, consisting of the six States of the Union lying to the eastward of that of New York ; and I may here add, that if we had continued our voyage for 30 hours longer,* winding along the coast, we might have reached the famous sea-port of New York, the Liverpool of the United States.

From Boston towards the south (a little easterly), 36 miles distant, is the town of *Plymouth*, near which the first settlers landed on the 22d of December, 1620; and from Boston, towards the north-west, 26 miles distant, is the city of *Lowell*, which constitutes the termination of our present travels.

The city of Lowell, to which we have now brought you, is situated on the river Merrimack, proximate to the junction of the rivers Merrimack and Concord. The ample

* The distance from the harbor of Boston to New York, coast-wise, is about 350 English miles; but the estimate of time here is for a continuation of the course from Halifax without touching at Boston.

supply of water, with the advantage of a fall of about 30 feet, was the chief encouragement, through the water-power thus afforded, to the establishing of manufactories here, and has proved the .occasion of their great extension and rapid increase, and of the consequent growth of the town.

SECT. 2. Something may be now said about *The history of Lowell, from its commencment to the present time*, for which a very short space will be sufficient.

Little more than twenty years ago, the place where Lowell stands, was but farm-land or wilderness. A small factory, indeed, was erected upon the Merrimack in the year 1813, being the first attempt to apply these favorable waters to the promotion of manufactures. So humble was this attempt, that the whole cost of the erection of the factory is stated to have been but 3000 dollars, or scarcely 600 guineas. But the great advantages of the water-power of the situation, for profitable enterprise, being fully determined, companies were formed—here called "corporations"—and their objects so rapidly promoted, that a large extent of ground soon became covered with buildings, and crowded with population.

The preparatory measure for rendering the waters of the rivers available for an extensive range of factories, was the formation of reservoirs, canals and locks, works which had long been contemplated, but not brought into effective operation, it would seem, until the year 1822. The year following (1823) the *Merrimack* corporation commenced operations for the spinning of cotton, &c., with a capital stock either originally consisting of, or subsequently raised to, two million of dollars. This was followed in 1825 by the *Hamilton* corporation, with a capital of one million of dollars. To these succeeded, in 1828, the *Appleton* and the *Lowell*, each with 600,000 dollars capital. The *Middlesex* was the next, commencing operations in 1832; the *Suffolk* and *Tremont* in 1832; the *Lawrence* in 1833–4; the *Boott* in 1836; and the *Massachusetts* in 1840. The capital stock of the last three corporations in the ultimate amounts, up to the year 1844, was very nearly four millions of dollars.

Thus betwixt the years 1822 and 1840, eleven companies, or corporations (including that of the "locks and canals") commenced practical operations; the several companies, for the most part, having been incorporated in previous years; in four of them, however, the activity of the managers was such that they appear to have commenced spinning in the very year of their incorporation.

Along with the various mills of the several corporations, from two to five having been built under each, sprung up a corresponding quantity of other structures for dwelling-houses, warehouses, shops, hotels, with churches and other public buildings; and to these were added a number of appropriate places for carrying on other manufactures besides cotton, with all the variety of minor erections, called for by the trade of the city, and for the accommodation of its population.

The advance in the number of the population, has, in consequence of the progress in manufacturing enterprise, been singularly rapid. So recently as the year 1813, the first impression seems to have been made on the original wilds. In 1830, the population was 6474; in 1840, 20,796 (having been trebled in ten years); and in 1844, it was estimated at 25,000.

SECT. 3. I have next to give you *An account of my personal visit to Lowell.*

As a considerable variety of objects in that visit occupied my attention, I shall notice,—

(1.) *The appearance of the city, the factory-girls, and the factories.*

Anxious to see as much of Lowell as my time would enable me, I obtained, through a very kind and attentive friend, Amos Lawrence, Esq., of Boston, a previous arrangement for my visit there. And, accompanied by another gentleman, who showed me much considerate kindness, William Appleton, Esq., himself a considerable proprietor in the works there, I obtained most fully my desired object, and visited this place, so interesting on account of its manufacturing system and operatives, with the greatest possible advantage.

We set out by railway, on the morning of the 26th of August, and arrived at the station at Lowell, soon after twelve o'clock. The distance by railroad, 26 miles, is accomplished in about an hour and ten minutes, or, as in our case, some few minutes more.

The railway, different from any other on which I had before travelled in the United States, consists, like ours, of a double line, and 'is durably constructed, being laid on foundations of granite.' Its cost was 1,834,893 dollars, equal to 70,573 dollars, or upwards of £14,000 a mile,— a very unusual expense for an American railway. It well answers the purpose, however, to the proprietors, having paid, so long ago as 1841, 8 per cent dividend.*

On entering Lowell, a stranger is naturally struck with

* Boston Almanac, 1843, p. 41.

the contrast presented by that place, to an English manufacturing town. Here, in Bradford, for example, every building is of stone, or brick, solid, substantial, with little of the freshness that might be looked for in so rapidly an increasing town; there, in Lowell, though the mills and boarding-houses are generally of brick, the chief part of the other buildings, houses, hotels, and even churches, are of wood, and nearly the whole as fresh-looking as if built within a year. Here, with us, every thing, externally, is discolored with smoke; buildings, streets and causeways, alike bearing a sooty covering; the mud of the streets in color and consistency like blackish grey paint, and the air of heaven darkened as by a dense cloud: there, nothing is discolored, neither houses, nor mills, nor trees; the red brick factories and boarding-houses, and the other edifices of wood painted in light colors, look as fresh as if just finished; the streets, dusty enough, indeed, (for a deep dry covering of dust was on them when I was in Lowell), were yet not black like ours; and the sky, unshrouded by smoke or cloud, was brilliant and clear, the sun darting down its unobstructed rays with dazzling and scorching power. Then the trees and plants which, with us, soon become dingy after foliage bursts out, and which in some places, as in the church-yard of St. Peter's, being ever leafless, appear like the black stumps of the forest in America, the residue and remnants of the fire — there, in Lowell, were fresh and flourishing; the trees, with which some streets or mill precincts were screened and adorned, being verdant in their summer's foliage, and the neatly kept gardens, here and there seen about the factories, being amply adorned with floral beauty. Hence, as to Lowell, large as it has grown, it is yet rural in its appearance, and, notwithstanding its being a city of factories, is yet fresh and cleanly.

Much of this striking contrast arises from the difference of the agency employed for the propelling of the machinery, being, with us, *steam*, derived from a strong smoke-yielding coal; but with the manufacturers of Lowell, *water*, from above the falls of the river. But besides this difference, in the sources of the power made use of in the factories, they have a further advantage in the fuel employed for domestic purposes—this being either wood or anthracite coal,—the coal of the eastern country, which burns without smoke. Still, however, the difference betwixt us is, undoubtedly, much greater than it might be. Were such care and attention given to the consuming of the smoke with us, as the matter deserves and the public ought to demand, there would be a vast improvement in the condition of the atmosphere of our large manufacturing towns. For whatever may be the difficulties of consuming the smoke of our factory furnaces, I will be bold to say, that they are by no means insurmountable, or indeed extraordinary. Nay more, that among the variety of inventions which have already been made public, there are several, which, with *due care and attention in feeding the furnaces*, would be sufficiently effective to produce a decided improvement, and more than one of them, to accomplish, almost fully, the desired purpose. The leading principle adopted in most of these inventions is essentially the same, being to cause the smoke given off by fresh coal to pass, in close contact, over the surface of the clear fire of half-burned fuel, and thus to be consumed. In some of the more scientific processes, fresh air is introduced by pipes, or otherwise, and sometimes in a heated state, at the posi tion in the furnace where the consumption of the smoke is designed to take place; and in one process (the ingenious one patented by Mr. Ivison) *steam* is thrown into the fore-part of the furnace, above the fuel, by means of a fan-shaped distributer, which, it is said, has a very extraordinary effect in causing the disappearance of the smoke.*

It may, however, be observed, that scarcely any of these inventions will be found really effective, without a systematic attention to the judicious application of fuel by the firemen. Almost the worst of the plans that have been proposed will be productive of considerable advantage with *such* attention, and the best of them will fail without it. The fire, indeed, cannot be kindled without the production and escape of smoke; nor can it well be maintained, without some smoke, where the inferior qualities of coal are used, but when once the fire is fully got up, it may, in most cases, be carried on throughout the day without any material annoyance.

Perhaps the most effective mode of securing the requisite attention to that object would be by making such an advance on the wages of the engine-tenders as to enable them to bear a moderate fine, for allowing the emission of smoke, having regard to the time of such emission, and so that the extra wages would amount to a *premium*, where the duty should be performed with proper diligence and attention. And the effect, I apprehend, would be abundantly adequate to repay any additional expenses.

But, returning from this digression, let me proceed with my account of what I saw during my visit to Lowell. Soon after arriving at the railway station at Lowell, we

* Encyclopædia Britannica, Ed. 7, Art. "Smoke."

were joined by a gentleman, Mr. Samuel Lawrence, a principal in a large woollen factory, who had been expecting us; and we had also the assistance of the manager of the cotton-mills of one of the principal corporations.

Having deposited my travelling bag at the Merrimack House, and ordered dinner for a little party at an hour convenient for the object in view, we proceeded, without further delay, to one of the factories, that we might see the factory-workers as they came out to their dinners. And, to a stranger, it was an interesting sight. Several hundreds of young women, but not any children, issued from the mills immediately within view, altogether very orderly in their manner, and very respectable in their appearance. They were neatly dressed, and clean in their persons; many with their hair nicely arranged, and not a few with it flowing in carefully curled ringlets. All wore (being the height of summer) a light calico-covered bonnet, or sort of *caleche*, large enough to screen the face, and with a dependent curtain shielding the neck and shoulders. Many wore veils and some carried silk parasols. By no means a few were exceedingly well-looking, more pallid than the factory-girls with us, and generally slight in their figure. There was not the slightest appearance of boldness or vulgarity; on the contrary, a very becoming propriety and respectability of manner, approaching, with some, to genteel.

They had not far to go for their refreshment, their boarding-houses, in which most of them were lodged, being generally contiguous to the factories wherein they were employed.

My young friends of the factories, whom I see before

me, will, I hope, here take a hint as to what is becoming in young women on leaving the mills for their meals. For when I sée the levity of behavior of some of our mill-girls as they come out from their work into the street, I cannot but wish that I could show them how their sisters in America conduct themselves, at least the many hundreds whom I saw at Lowell, under the same circumstances. And sure I am, that every one who knows the self-respect that ought to be entertained, or feels the *value of character* to a young female, would try to imitate the good and pleasing example to which I have referred. And let me tell you what I consider to be the fact, that no such thing is ever to be seen among the young women of Lowell, as *rudeness*, and very rarely indeed, I believe, any thing like levity or immodesty of behavior in the streets. Now do you, my young friends, keep this in your minds, not only as a rule for your own conduct, but as a check to other girls in case of rudeness or immodesty of conduct in your presence. For, remember, that whilst a modest, obliging, pure-minded and well-principled young person always meets with respect, and, with those of her acquaintance whose good feelings are the most valuable and worthy of being cherished, with affectionate consideration, — a bold, rude, vulgar and immodest girl is a disgusting creature, and a disgrace to her sex. Speaking, as you are aware I now do, in kindly and pastoral affection, you will receive my remarks, I hope, in the like spirit. My earnest desire is to be useful to you; if therefore you will join me in the effort, by the blessing of God, we shall get good.

One observation more, I will take this opportunity of making, on the practice so common here of our factory-girls, or at least many of them, going into the streets with their heads uncovered. This you may think a matter oi no consequence, neither, perhaps is it, in itself, of any very great consequence. Yet, let me tell you, that strangers have observed it, and considered it an indecorum — some, indeed, having supposed that young persons going in the streets without bonnet or cap or sufficient covering could hardly be modest girls. I would not allow myself, however, to draw an inference to this extent of importance from the habit referred to; but still I will remind you, that the circumstance of a woman appearing in public with her *head uncovered* has not been thought unworthy of being mentioned, and rebuked, in that Holy Book which is the only sure guide to what is becoming as well as to what is right.*

The mills, in their general appearance, (which the interval of dinner-time enabled me more particularly to notice) much resemble ours, except that they generally looked smaller than our Bradford factories, and are built of brick, instead of stone. They constitute an extensive range near the river side, but so planted as to receive their share of water, as conveniently as may be, from the different canals by which water-power is supplied. Generally, I think, they are from four to five stories high, besides a floor, in many of them, in the central part of the roof. There were twenty windows in a row, in the front of one of the mills of four stories which I more specially noticed.

Long ladders, attached to the outside of the buildings, to some not less than four or six, extending from the ground to the roof, particularly excited my attention. These were

* 1 Cor. xi. 5, 6.

fire escapes, so arranged, both on the sides and at the ends of some of the buildings, as to afford the means of escaping outside, independent of the stairs (which are of wood), from every floor, and almost from every subdivision of the factory. In some cases the ladders ran quite over the roof. The object of this might be to facilitate measures for extinguishing fire; an object, indeed, which the ladders generally would also subserve.

Every thing about the mills is characterized by cleanliness and neatness. Nice, tastefully laid out gardens or plots of grass, with trees and shrubs, adorn the enclosures of many of the factories; and every thing is strikingly fresh and fair.

2. *Visit to a Cotton Factory.*

On the recommencement of work, after dinner, we went into, and all through, the adjoining cotton-factory. The interior, except in the departments used for the first steps of their operations, which are necessarily dusty, corresponded in neatness and cleanliness with what the outside had indicated. The floors and staircases were beautifully clean, and the stairs painted, except a space, as if *meant* for a stair-carpet, in the middle. The windows were ornamented, in some of the departments, with plants set in flower-pots, the property of the girls, who seemed to have a taste for their cultivation, and a pride in keeping them nicely.

The carding, spinning, and weaving of cotton were all carried on in the same building. About 250 hands were employed, perhaps the average number,* in the first mill

* They range from about 200 to 300 hands.

I examined, but a woollen mill of Mr. Lawrence's had near 800.

The young women observed at work in this cotton-mill were all above the age of children; I supposed, generally, they were from fifteen to twenty, though many were much older; and some few, I believe, were married women. I afterwards learnt that the average age was about twenty-three years. The great majority were from the country; but on this point, as well as about that of the case of children, I shall have to speak hereafter.

On almost every floor of the factory, I conversed with one or more of the females at work. In all cases their manner was easy and unaffected, modest and respectful. In many of them there was an intelligent, and rather refined expression of countenance, characterized, in some, with a cast of thoughtfulness.

When you see a young person, under such circumstances, appearing very thoughtful, any one, that was familiarly intimate with her, would very naturally enquire what she was thinking about? Perhaps some of the factory workers here this evening, if they had seen a serious and intelligent looking girl as I did, appearing very thoughtful, would have been curious to know something about her, or about the current of her thoughts? If there be, let me tell you that it may not be impossible to satisfy your curiosity. Not, indeed, as to what such particular girl was at that time thinking about; but what one or other of the factory girls, similarly situated, had some time back been thinking about.

We have the results of the thoughts of many of them printed — yes, printed! — in the books I hold in my hand, which, perhaps some of you have seen, and, if not, which you ought to see. They are parts or numbers of a periodical work, written entirely by the factory-girls of the place I visited. It is called the "Lowell Offering." Mr. Knight, an enterprising London publisher, has placed some of the most interesting matter of the original work within your reach. In this little volume, entitled "Mind among the Spindles," you have a cheap and very judicious selection of the papers.

One of them more particularly serves my present purpose, and enables me to tell you something of the thoughts that have passed through the mind of a Lowell factory-girl. It is called "A Weaver's Reverie." Let me read it to you: and, if it be new to you, I will venture to predict that it will surprise you — perhaps not a little.

A WEAVER'S REVERIE.

It was a sunny day, and I left for a few moments the circumscribed spot which is my appointed place of labor, that I might look from an adjoining window upon the bright loveliness of nature. Yes, it was a sunny day; but for many days before, the sky had been veiled in gloomy clouds; and joyous indeed was it to look up into that blue vault, and see it unobscured by its sombre screen; and my heart fluttered, like a prisoned bird, with its painful longings for an unchecked flight amidst the beautiful creation around me.

Why is it, said a friend to me one day, that the factory-girls write so much about the beauties of nature?

Oh! why is it, (thought I when the query afterwards recurred to me,) why is it that visions of thrilling loveliness so often bless the sightless orbs of those whose eyes have once been blessed with the power of vision?

Why is it that the delirious dreams of the famine stricken, are of tables loaded with the richest viands, or groves, whose pendent boughs droop with their delicious burdens of luscious fruit?

Why is it that haunting tones of sweetest melody come to us in the deep stillness of midnight, when the thousand tongues of man and nature are for a season mute?

Why is it that the desert-traveller looks forward upon the burning boundless waste, and sees pictured before his aching eyes, some verdant oasis, with its murmuring streams, its gushing founts, and shadowy groves — but as he presses on with faltering step, the bright *mirage* recedes, until he lies down to die of weariness upon the scorching sands, with that isle of loveliness before him?

Oh tell me why is this, and I will tell why the factory-girl sits in the hour of meditation, and thinks — not of the crowded clattering mill, nor of the noisy tenement which is her home, nor of the thronged and busy street which she may sometimes tread, — but of the still and lovely scenes, which, in by-gone hours, have sent their pure and elevating influence with a thrilling sweep across the strings of the spirit-harp, and then awakened its sweetest, loftiest notes; and ever as she sits in silence and seclusion, endeavoring to draw from that many-toned instrument a strain which may be meet for another's ear! that music comes to the eager listener like the sound with which the sea-shell echoes the roar of what was once its watery home. All her best and holiest thoughts are linked with those bright pictures which called them forth, and when she would embody them for the instruction of others, she does it by a delineation of those scenes which have quickened and purified her own mind.

It was this love of nature's beauties, and a yearning for the pure hallowed feelings which those beauties had been wont to call up from their hidden springs in the depths of the soul, to bear away upon their swelling tide the corruption which had gathered, and I feared, might settle there, — it was this love, and longing, and fear, which made my heart throb quickly, as I sent forth a momentary glance from the factory window.

I think I said there was a cloudless sky; but it was not so. It was clear and soft, and its beauteous hue was of "the hyacinth's deep blue" — but there was one bright solitary cloud, far up in the cerulean vault; and I wished that it might for once be in my power to lie down upon that white, fleecy couch, and there, away and alone, to dream of all things holy, calm, and beautiful. Methought that better feelings, and clearer thoughts than are often wont to visit me, would there take undisturbed possession of my soul.

And might I not be there, and send my unobstructed glance into the depths of ether above me, and forget for a little while that I had ever been a foolish, wayward, guilty child of earth? Could I not then cast aside the burden of error and sin which must ever depress me here, and with the maturity of womanhood, feel also the innocence of infancy? And with that sense of purity and perfection, there would necessarily be mingled a feeling of sweet uncloying bliss — such as imagination may conceive, but which seldom pervades and sanctifies the earthly heart. Might I not look down from my aerial position, and view this little world, and its hills, valleys, plains and streamlets, and its thousands of busy inhabitants, and see how puerile and unsatisfactory it would look to one so totally disconnected from it? Yes, there, upon that soft snowy cloud could I sit, and gaze upon my native earth, and feel how empty and "vain are all things here below."

But not motionless would I stay upon that aerial couch. I would call upon the breezes to waft me away over the broad blue ocean, and with nought but the clear bright ether above me, have nought but a boundless, sparkling watery expanse below me. Then I would look down upon the vessels pursuing their different courses across the bright waters; and as I watched their toilsome progress, I should feel how blessed a thing it is to be where no impediment of wind or wave might obstruct my onward way.

But when the beams of a mid-day sun had ceased to flash from the foaming sea, I should wish my cloud to bear away to the western sky, and divesting itself of its snowy whiteness, stand there, arrayed in the brilliant hues of the setting sun. Yes, well should I love to be stationed there, and see it catch those parting rays, and, transforming them to dyes of purple and crimson, shine forth in its evening vestment, with a border of brightest gold. Then could I watch the king of day as he sinks into his watery bed, leaving behind a line of crimson light to mark the path which led him to his place of rest.

Yet once, O only once, should I love to have that cloud pass on — on — on — among the myriads of stars; and leaving them all behind, go far away into the empty void of space beyond. I should love, for once, to be *alone*. Alone! where *could* I be alone? But I would fain be where there is no other, save the INVISIBLE, and there, where not even one distant star should send its feeble rays to tell of a universe beyond, there would I rest upon that soft light cloud, and with a fathomless depth below me, and a measureless waste above and around me, there would I ———

"Your looms are going without filling," said a loud voice at my elbow; so I ran as fast as possible, and changed my shuttles.

ELLA.

Take another instance, — the cogitations also of a factory worker, on observations she had made in the factory respecting the work and web of a little insect, which first produces the *materials* out of which the web is formed, works its wonderful machinery, within a space the size of a pin's head, without either steam or water, and then both spins and weaves without spindle or loom! You will have guessed, I suppose, that the little spinner and weaver I refer to is the *spider*.

The article is called "Cleaning-up;" being short, I shall read the whole.

CLEANING UP.

There is something to me very interesting in observing the manifestations of animal instinct — that unerring prompter which guides its willing disciple into the ever straight path, and shows him, with unfailing sagacity, the easiest and most correct method of accomplishing each necessary design.

But to enter here upon a philosophical dissertation, respecting the nature and developments of instinct, is not my design, and I will now detain you with but one or two instances of it, which have fallen under my own observation.

One warm day in the early spring, I observed a spider very busily engaged upon a dirty old web, which had for a long time curtained a pane of my factory window. Where Madam Arachne had kept herself during the winter, —

Really these factory girls are too learned; they employ such allusions as to send the experienced and the learned, who have forgotten their youthful studies, to their "classical dictionary" to ascertain what they mean! So in the case of the title of Madam *Arachne*, here given to the spider, we are reminded of the heathen-classic story told of Arachne, a woman of Colophon, who was so expert with her needle that she challenged Minerva (the goddess of wisdom and the liberal arts) to a trial of skill. But though she accomplished wonders, when the trial came, she was defeated by the goddess, and hanged herself in despair; but the goddess, as the story goes, restored her life, changing her into a spider!

So much for the explanation; let us proceed with the description before us:—

Where Madam Arachne had kept herself during the winter, was not in my power to ascertain; but she was in a very good condition, plump, spry, and full of energy. The activity of her movements awakened my curiosity, and I watched with much interest the commotion in the old dwelling, or rather slaughter-house, — for I doubted not that many a green-head and blue-bottle had there met an untimely end.

I soon found that madam was laboriously engaged in that necessary part of household exercises called CLEANING UP; (in some parts of Yorkshire, it is called, I believe, "thorough cleaning,") and she had chosen precisely the season for her labors which all good housewives have by common consent appropriated to painting, cleaning, white-washing, &c. With much labor, and a prodigal expenditure of steps, she removed, one by one, the tiny bits of dirt, sand, &c., &c., which had accumulated in this net during the winter; but it was not done, as I at first thought, by pushing and poking, and thrusting the intruders out, but by gradually destroying their *location*, as a western emigrant would say. Whether this was done, as I at one time imagined, by devouring the fibre, as she passed over it, or by winding it around some under part of her body, or whether she left it at the centre of the web, to which point she invariably returned after every peregrination to the outskirts, I could not satisfy myself. It was to me a cause of great marvel, and awakened my perceptive as well as reflective faculties from a long winter nap.

To the first theory there was no objection, excepting that I had never heard of its being done; but then it might be so, and in this case *I had discovered* what had escaped the observation of all preceding naturalists. To the second there was this objection, that when I occasionally caught a front view of "my lady," she showed no distaff, upon which she might have re-wound her unravelled thread. The third suggestion was also objectionable, because, though the centre looked somewhat thicker, or I surmised that it did, yet it was not so much so as it must have been had it been the depot of the whole concern.

Of one thing I was at length assured; that there was to be an entire demolition of the whole fabric, with the exception of the main beams (or sleepers, I think, is the technical term,) which remained as usual, when all else had been removed. Then I went away for the night, and when I returned the next morning, expecting to behold a blank — a void — an evacuation of premises — a removal — a disappearance — a destruction most complete, without even a wreck left behind — lo! there was again the rebuilt mansion — the restored fabric — the reversed Penelopian labor; and madam was rejoicing like the patient man of Uz, when more than he had lost was restored to him.

My feelings (for I have a large bump of sympathy) were of that pleasurable kind which Jack must have experienced when he saw the castle which in a single night had established itself upon the top of his bean-pole; or which enlivened the bosom of Aladdin, when he saw the beautiful palace which in a night had travelled from the genii's dominions to the waste field which it then beautified; and I felt truly rejoiced that my industrious neighbor's works of darkness were not always deeds of evil. But alack for the poor *spinster*, when it came *my* turn to be *cleaning up!* M. E.

(3.) *A Carpet Manufactory, &c.* — We next visited a *carpet manufactory*, where the best patterns, and various qualities of the Kidderminster kind, are made. Here, for the first time, I witnessed the weaving of figured patterns in carpets, in a variety of colors, by the Jacquard loom. As every color in the weft must have its own shuttle, and as the colors are perpetually varying, the ingenuity of the invention consisted in bringing the shuttles of the proper color opposite the opening of the web in quick and proper succession. This is accomplished by fixing the shuttle-guide in a moveable frame containing many stages or tiers of shuttles, which frame being raised or let down by the machinery to the level of the proper color for the weft, allows the stroke on the shuttle to send that color through the web.

The effect of this invention on the cost of producing the carpet was, on its first introduction, $8\frac{1}{2}$d. per yard; the cost of weaving being reduced from 24 cents to 7 cents; that is, so many half-pennies, two cents being equal to an English penny. A young woman attending one of these new looms, I was told, being paid 7 cents per yard of carpeting completed, was then earning, for 15 yards made in a day, 105 cents, or 4s. $4\frac{1}{2}$d. per day! This was, however, only what might be called an experimental price; as they would hardly continue to pay for the superintending of these looms, after they had become common in their manufactory, at so high a rate.

A large *Print-Work* we also visited, first passing through the scouring, bleaching, and dying departments, and afterwards proceeding to the floors where the calicoes were printed, steam-dried, and subsequently pressed or callendered, and packed for the market. Of the extent of these various works, and their several products, I shall have some account to give at the conclusion of my observations; but I may just notice, as to this particular print-work, that 1100 pieces or about 30,000 yards of calicoes (I hope I remember the numbers correctly) are here bleached, dyed, printed, and issued in a finished state, daily.

Besides these works, we just looked into some other manufactories; hastily inspected a machinery-making establishment; went through a second cotton-factory, and conversed with more of the persons employed therein; and then, not a little jaded by the efforts of climbing stairs and walking from place to place in a hot, oppressive day, we went to our hotel and refreshed ourselves with our dinner.

After dinner we again moved abroad, renewing, for a short time, the survey I designed of the works carried on in the city, and finishing this part of the business by a visit to one of the lodging or boarding-houses of the factory-girls, and by a drive out into the suburb, to the hospital for the factory sick.

(4.) *The Corporation Boarding-houses.*

As few of the young women have their parents residing in Lowell—most of them coming up from the country for the sake of the employment for themselves—they live chiefly, but not entirely, in *boarding-houses* expressly built by the corporations for their accommodation. These houses are neatly built of brick, in rows or in short streets broken up into what, in America, are called "blocks." Usually, they are three stories high, with a garret, and contain six or seven bed-rooms with two beds, calculated for two persons each, so that the whole will duly accommodate from twenty-four to thirty girls, and some, I believe, more, when full.

The one I visited was not *selected*, but taken at a venture. It was in the evening, just as the young women were leaving their work. I was shown through almost every room, from the attic to the ground floor. Everything was clean and orderly, presenting a comfortable appearance. The rooms were better than I had often had at some of the hotels; the beds of a commodious size, and the furniture sufficiently good and neat. Some of the young people had just come in, and two were in the parlor, with whom (one of them more particularly) I sat and conversed for some time. She was tall in person, of slight figure, and rather tastefully dressed, (I think in a muslin dress) with a *gold* pencil case suspended by a silk cord round the neck. She conversed freely, in good language, and without either affectation or restraint. I told them that I came from a large manufacturing town in England, and that I was wishful to see the American factories and factory-workers, that I might tell my parishioners something about them. This intimation, indeed, was calculated, *perhaps* to interest them, but *certainly* to put them on their guard. But I did not observe that it produced any check to free and frank communication.

In this place, I might just notice, that their manner of living in their boarding-houses is amply liberal for the payments they make. They have three meals a day, combining tea or coffee, at each, along with substantial provision; and differing but little, in the variety, I believe, except that hot meat is the prevalent advantage given to the dinner. The two systems of boarding-houses, that of the corporations and that of private arrangement, serve as a check upon each other; so that if any abuse were to creep into either, or if the quality or quantity of provision, or the amount charged, should be unsatisfactory in the one class, the inmates would, in most cases, (except where no private lodging-houses are situated near their mills) have the opportunity of removing their residence. For though some of the corporations require their female operatives to reside in the houses built and owned by them, and most of them consider it right and proper that they should do so, yet, where the mills happen to be situated in the midst of private houses, large numbers of the hands sometimes board away from the corporations.*

As to expense of board and lodging, the charge limited by the several corporations for the female factory operatives, in their own houses, is so exceedingly moderate, being, I believe, only a dollar and a quarter, or about five shillings and two-pence, English, a week, that no system of imposition can possibly be carried on therein, or in the independent lodging-houses elsewhere.

* Dr. Bartlett's Vindication, &c., p. 7.

(5.) The *Hospital* for the factory-workers — the last establishment of a public nature which I visited — is an admirable institution. It is beautifully situated, on an elevated site in a suburb of the city. Originally, it was a private gentleman's residence ; the best, or one of the best houses in the neighborhood, and was purchased by the corporations for its present purpose at an expense, if I recollect rightly, of 15,000 dollars.* The patients, or sick boarders, assist considerably in the expense of maintaining the establishment — which is admirably superintended by Dr. G. Kimball and his wife — by a weekly contribution of three dollars, equal to twelve shillings and sixpence, English, for board, attendance, advice, &c. But this charge has reference to their ability to pay, for no factory-worker is ever excluded from the benefits of the institution because of being *unable* to pay. We went through most of the *rooms* — for the building retains its divisions into rooms instead of wards — which were sufficiently commodious for two to four beds, nicely furnished, and with much appearance of comfort. There were but few inmates ; with some of them I stopped a few moments for conversation, and to suggest some serious thought, on the dispensation of a heavenly Father's hand, which had required their availing themselves of this advantageous refuge. Dr. Kimball, to whom I had a letter of introduction from a mutual professional friend in Cincinnati, was unfortunately from home. Mrs. Kimball, however, kindly showed us every possible attention.

The examination of this place of airy and commodious retreat for the sick, was very satisfactory and interesting. For those young persons, far estranged from their parental

* The first cost was nearly 30,000.

home, and a parent's watchful affectionate care, the best practicable compensation seems here to be provided. With the most skilful medical attendance and advice, is combined a peculiar consideration, for it was evident that the patients were cared for, as if constituting a part of the family and household of the kind and intelligent superintendents.

After this visit, I was driven into the country about a couple of miles, to the residence of Mr. Samuel Lawrence, — the gentleman to whose kind offices I had been specially committed, — the friend who had accompanied me from Boston having returned home. Here I took tea, along with the family and a party of friends residing in the house, together with an Episcopal clergyman who had been invited to meet me, and from whom I obtained some interesting and useful information.

Being anxious to see the factory operatives at their leisure and recreation in the evening, I left very early, and being conveyed to my hotel, in the centre of the city, I went from thence through the principal streets, among which — the evening being cool and agreeable — I pleasantly rambled for upwards of an hour. Extensive ranges of well-furnished shops, brilliantly lighted up, and continuing so until nine o'clock, or a little later, indicated the prosperity of the place, whilst a very large number of the young women of the factories were seen enjoying the evening air in the streets, and not a few availing themselves of the facilities offered them for shopping. Comparatively few men were about. The girls were generally walking two or more together, and I did not observe the smallest sign of levity or indecorum in any one individual, amongst the many hundreds whom I passed.

Having purchased a few numbers of the "Lowell Offering," I retired early to my room in the Merrimack House and put down a series of notes, from which, aided by the recollections of a day spent in so interesting a manner, and some official documents, the substance of this lecture has been mainly gathered.

In the morning, before the starting of the railway cars, by which I had arranged to return to Boston, so as to be in time to go out to Cambridge to attend the first of the "Commencement" meetings of Harvard University, (a Collegiate Institution of more than 200 years standing,) I made an additional short examination of the streets in the vicinity of the railway station. The results of this examination are incorporated with other parts of my present Lecture.

SECT. 4. *Wages of the Factory Operatives.*

The subject of wages must be one of so much interest with both laborers and employers, that I should hardly make my Lectures complete without referring to it.

Wages in America are generally good, in many cases high. Yet for mere physical labor, the use of muscles and sinews only, requiring little intellect or experience or handicraft cleverness, the advantage in America is not so great as is commonly supposed.

But not to detain you by speaking about wages generally, I shall proceed to give you some information about the earnings of factory operatives.

The average wages of the *males* employed in the Lowell factories, amount, on the authority of the table of "Statistics of Lowell Manufactures," to about 70 cents (2s. 11d.) por *day*, clear of board; and of the *girls* to a dollar and three quarters (about 7s. 3d.) per *week*, clear of board and lodging. But Dr. Bartlett, of Lowell, in an interesting pamphlet entitled "A Vindication of the character and condition of the females employed in the Lowell Mills," rates it at about two dollars, or 8s. 4d. Those, therefore, who are at all economical, may easily lay by, on an average, three or four shillings weekly, thus accumulating a saving of £8 to £10 a year; whilst those who are clever operatives may earn, in some of the departments of the mills, twice the wages of the general average, and of course may save, if so disposed, a much greater proportion, possibly to the amount of from twenty to thirty pounds a year. Some of the weavers — who, being clever, are able to attend three or four looms — may, in prosperous times, realize, I believe, the higher amount, if not more. But as clothing is expensive, and the girls dress very gay, these large *possible* savings are probably confined to but few.

In a letter by "a Lowell Factory Girl," which appeared in a Boston paper of Nov. 9, 1844, some curious information was given respecting her own earnings, as a weaver, under different circumstances in regard to speed, wages, &c. during the years 1842, 1843 and 1844. As some of you may be interested in it, I will give you the extract, without abridgment, containing these particulars. Just premising that she speaks from experience in her own case solely, as to this matter, she says: —

"In May, 1842, the last month before the reduction of wages, I tended two looms, running at the rate of 140 beats of the lathe per minute. In twenty-four days I earned 14 dollars 52 cents. In the next month, June, when speed and prices had both been reduced, I tended four looms, at a speed of 100, and earned in 24 days, 13 dollars 52 cents, and I certainly, after the first few days, had an easier task than with two looms at the high speed. I increased my earnings every month a little, by the gradual increase of the speed, as I grew accustomed to it. In January, 1843, the speed was raised to about 118, and the price reduced still lower. I earned in that month, in 24 days on three looms, 14 dollars 60 cents, and my work was in no degree harder. The speed was raised just as we could bear it, and often, almost always, at our own request, because with the increase of speed our pay increased. In June, 1843, I still tended three looms, and in 24 days earned 15 dollars 40 cents, and in June, 1844, feeling able to tend four looms at a speed of about 120, I received 16 dollars 92 cents (equal to £3. 10s. 6d.) in payment for 24 days' work. I affirm that I have not in any of these, or other months, overworked myself. I have kept gaining in ability and skill, and as fast as I did so I was allowed to make more and more money, by the accommodation of the speed of the looms to my capacity. I am by no means the best weaver in the room where I work, though perhaps better than the average. I believe I have given no exaggerated picture of what has been the true average of girls. The other departments I suppose to have fared much as we in the weaving rooms."

This amount of earnings by a young woman at Lowell, you will remember, is considerably beyond the average, that being about three dollars per week, (12s. 6d.) *not* deducting expense of board. And the difference of the wages in America and this country arises, mainly, from one simple fact, that there, labor being scarce, it is necessarily dear; whilst here it is generally too plentiful, and consequently cheap.

Sect. 5. *Statistics of Lowell at the present time.*

It yet remains that I should give you a summary view of the statistics of this interesting and greatly thriving city. And that you may duly estimate the nature of its progress, you will please bear in mind the fact already stated, that it is almost entirely the growth of a period of about two and twenty years.

The corporations or companies for the manufacture of

cotton and woollen goods, but chiefly of cotton, which is the staple manufacture, are ten in number, these corporations being the proprietors of 32 mills, besides print-works and other manufacturing concerns. Their total capital is 10,650,000 dollars, or about £2,200,000 sterling. The total number of hands employed by these several companies at the commencement of last year (1844) was stated to be 6295 females, and 2345 males; but in August last I was told, that the number of girls and women was about 7000. The whole of the spinning works run 201,076 spindles for cotton and wool. The number of looms was 6265, at the time mentioned; but they have since, I believe, been considerably increased.

The rate of production was, then, 1,425,800 yards of cloth weekly, or 74,141,600 yards annually. Nearly twenty-three millions of pounds of cotton were wrought up or consumed in a year, 100 pounds of cotton producing, in their fabrics, 89 pounds of cloth. The medium produce of a loom, in weaving of thread or yarn No. 14, is forty-four to forty-five yards a day, and of No. 30, thirty yards. I believe those gentlemen present, who are acquainted with the results of our power-weaving, would consider that the Lowell looms are worked at a great speed.

Of the texture of calicoes, &c. here produced, it will be found that each spindle yields sufficient yarn for a yard and one-tenth per day.

The principal kinds of goods manufactured at Lowell, commencing with the largest quantities, are, sheetings, shirtings, printed cloths, of Numbers 14 to 40, drillings, carpetings, blankets, broad cloths, cassimeres, &c.

The production in dyed and printed cottons, was, in January 1844, 273,000 yards weekly.

The factories are heated with steam, or hot air, for which object, and other purposes connected with the factories and print-works, chiefly, 12,500 tons of anthracite coal are used annually, together with 3290 cords of wood, a cord being a compact pile of 8 feet by 4 and 4.

The oil required for the machinery and wool, &c., amounts to 67,849 gallons a year.

The diameter of the water-wheels made use of for working the machinery, varies from 13 to 30 feet; and the "length" of wheels, or width (?), for each mill, varies from 21 to 60 feet.

In giving these results and statistics concerning Lowell, I am well aware, that, in many particulars, they fall far short of the work carried on in some of our large manufacturing towns in England and Scotland. But still, I thought they would be interesting to some of you who are engaged in the management, either as spinners, or merchants, of the staple business of our own enterprising and rapidly advancing town.

To the principal establishments already enumerated may be added, the Lowell water-proofing concern; the extensive powder mills; the "Lowell bleachery;" flannel mill; batting mill; paper mill; card and whip factory; grist and saw mills, &c. &c., together employing about 500 hands, and a capital of half a million of dollars.

"The locks and canal company," which supplies the water, and arranges its canals and branch aqueducts, also takes the work, I believe, of erecting mills, and constructing the machinery. In times of active operations it employs from 1000 to 1200 hands.*

* Extracted from "Statistics of Lowell, Jan. 1, 1844, compiled from authentic sources," and put into my hands by the manager of one of the corporations.

As to the statistics of population, I need only mention, in addition to what has been stated, a remarkable circumstance arising out of the influx of young women coming *alone* from the country; that of the 25,000 persons, the estimated population of the city, more than one-half are represented as being betwixt the ages of fifteen and thirty years.

Of other statistics, I may just name, that there are no less than twenty-two places of public worship, comprising the ample variety of ten different denominations, in this youthful city. There are Sunday Schools and daily schools in full measure, I believe, proportionate to the requirements of the youthful population. And besides all the varieties of literary and mutual improvement institutions, and the libraries for general reading; there is a supply of periodicals, of the newspaper class, which, to an Englishman appears quite curious; though to an American, the variety, in such publications, would not be deemed extraordinary. Thus Lowell, for the information or amusement of its 25,000 of total population, produces one daily newspaper, two thrice a week, and six weekly, besides the clever periodical conducted by the factory-girls, the Lowell Offering, which is a monthly publication! In the whole, nine newspapers,—some of them, I apprehend small in quantity, whatever may be the quality, yet constituting a supply for the reading of the masses, which presents a curious contrast to the public press in some parts of this country; in Bradford, for instance, where we have one newspaper, with a second edition weekly for 60,000 population in the town and its precincts, and 120,000, perhaps, in the parish!

So far as this portion of my subject, designed for this evening's consideration, goes, I might here bring my present Lecture to a close. Before I do so, however, I should wish to interest you somewhat further in the female factory operatives of Lowell, so that we might enter on our next Lecture with a more favorable prepossession respecting them, when I seek to stimulate you to emulate them, in whatever may be found elevating or superior in their example. For this purpose, as well as really for giving more interest to my own Lecture, I shall read to you, in conclusion, some further extracts from their own writings, as published in the Lowell Offering. These, for the present occasion, I may advantageously take from Mr. Knight's judicious selection, which I have already commended to your attention.

The first article, which I have chosen for my present purpose, is entitled, "The Indian Pledge," which, being a story of only two or three pages, I may give you entire.

THE INDIAN PLEDGE.

On the door-steps of a cottage in the land of "steady habits," some ninety or a hundred years since, might, on a soft evening in June, have been seen a sturdy young farmer, preparing his scythes for the coming hay-making season. So intent was he upon his work that he heeded not the approach of a tall Indian, accoutred for a hunting expedition, until, "Will you give an unfortunate hunter some supper and lodging for the night?" in a tone of supplication caught his ear.

The farmer raised his eyes from his work, and darting fury from beneath a pair of shaggy eyebrows, he exclaimed, "Heathen, Indian dog, begone! you shall have nothing here."

"But I am very hungry," said the Indian; "give only a crust of bread and a bone to strengthen me, on my journey."

"Get you gone, you heathen dog!" said the farmer; "I have nothing for you."

"Give me but a cup of cold water," said the Indian, "for I am very faint."

This appeal was not more successful than the others. Reiterated abuse, and to be told to drink when he came to a river, was all he could obtain from one who bore the name of Christian! But the supplicating appeal fell not unheeded on the ear of one of finer mould and more sensibility. The farmer's youthful bride heard the whole, as she sat hushing her infant to rest; and from the open casement she watched the poor Indian until she saw his dusky form sink, apparently exhausted, on the ground, at no great distance from her dwelling. Ascertaining that her husband was too busied with his work to notice her, she was soon at the Indian's side, with a pitcher of milk and a napkin filled with bread and cheese. "Will my red brother slake his thirst with some milk?" said this angel of mercy; and as he essayed to comply with her invitation, she untied the napkin, and bade him eat and be refreshed.

"Cantantowwit protect the white dove from the pounces of the eagle," said the Indian; for *her* sake the unfledged young shall be safe in their nest, and her red brother will not seek to be revenged."

He then drew a bunch of feathers from his bosom, and plucking one of the longest, gave it to her, and said, "When the white dove's mate flies over the Indians' hunting grounds, bid him wear this on his head."

* * * * * * * * *

The summer had passed away. Harvest time had come and gone, and preparations had been made for a hunting excursion by the neighbors. Our young farmer was to be one of the party; but on the eve of their departure he had strange misgivings relative to his safety. No doubt his imagination was haunted by the form of the Indian, whom, in the preceding summer, he had treated so harshly.

The morning that witnessed the departure of the hunters was one of surpassing beauty. Not a clould was to be seen, save one that gathered on the brow of Ichabod (our young farmer), as he attempted to tear a feather from his hunting cap, which was sewed fast to it. His wife arrested his hand, while she whispered in his ear, and a slight quiver agitated his lips as he said, "Well, Mary, if you think this feather will protect me from the arrows of the red skins, I'll e'en let it remain." Ichabod donned his cap, shouldered his rifle, and the hunters were soon on their way in quest of game.

The day wore away as was usual with people on a like excursion; and at nightfall they took shelter in the den of a bear, whose flesh served for supper, and whose skin spread on Bruin's bed of leaves, pillowed their heads through a long November night.

With the first dawn of morning the hunters left their rude shelter and resumed their chase. Ichabod, by some mishap, soon separated from his companions, and in trying to join them got bewildered. He wandered all day in the forest, and just as the sun was receding from sight, and he was about sinking down in despair, he espied an Indian hut. With mingled emotions of hope and fear, he bent his steps towards it; and meeting an Indian at the door, he asked him to direct him to the nearest white settlement.

"If the weary hunter will rest till morning, the eagle will show him the way to the nest of his white dove," said the Indian, as he took Ichabod by the hand and led him within his hut. The Indian gave him a supper of parched corn and venison, and spread the skins of animals which he had taken in hunting, for his bed.

The light had hardly begun to streak the east, when the Indian awoke Ichabod, and after a slight repast, the twain started for the settlement of the whites. Late in the afternoon, as they emerged from a thick wood, Ichabod with joy espied his home. A heartfelt ejaculation had scarce escaped his lips, when the Indian stepped before him, and turning around, stared him full in the face, and inquired if he had any recollection of a previous acquaintance with his red brother. Upon being answered in the negative, the Indian said, "Five moons ago, when I was faint and weary, you called me an Indian dog, and drove me from your door. I might now be revenged; but Cantantowwit bids me tell you to go home; and hereafter, when you see a red man in need of kindness, do to him as you have been done by. Farewell."

The Indian having said this turned upon his heel, and was soon out of sight. Ichabod was abashed. He went home purified in heart, having learnt a lesson of Christianity from an untutored savage. TABITHA.

My concluding selection will be taken from a paper, written also, remember, by a factory-girl, headed, "The Village Chronicle." This story I have selected not because there is much in it worth remembering, or calculated to instruct, except indeed as it shows how a sister's affection may be tenderly felt by one in a distant region, and how things which at home are little thought of may be deeply interesting to an absent brother, but because of its cleverness, and the simple nature evinced in the descriptions of the manner and feelings of two young persons in reading, under different circumstances, an ordinary village newspaper. It is first read, not very graciously or willingly, by Lina, a smart little girl of about twelve years of age, to her father, Mr. Wheeler; and then being forwarded, by an accidental opportunity, to Lina's brother, residing alone in the wilds of Kentucky, is read by him. The contrast in the feelings of the readers is as natural as it is characteristic. Though I shall be obliged in a degree to abridge it, sufficient of the spirit of it will, I hope, be preserved, to enable you to appreciate and enjoy it.

THE VILLAGE CHRONICLE.

"Come, Lina dear," said Mr. Wheeler to his little daughter, "lay by your knitting, if you please, and read me the paper."

"What, pa, this old paper, 'The Village Chronicle?'"

"Old, Lina! — why it is damp from the press. Not so old, by more than a dozen years, as you are."

"But, pa, the *news* is *old*. Our village mysteries are all worn threadbare by the gossiping old maids before the printer can get them in type; and the foreign information is more quickly obtained from other sources. And, pa, I wish you wouldn't call me Lina, it sounds so childish, and I begin to think myself quite a young lady, almost in my teens, you know; and Angeline is not so very long."

"Well, Angeline, as you please; but see if there is not something in the paper."

"Oh, yes, pa; to please you, I will read the stupid old (*new*, I mean) concern. Well, in the first place, we have some poetry; some of our village poets' (genius, you know, admits not of distinction of sex) effusions, or rather confusions. Miss Helena (it used to be Ellen once) Carrol's sublime sentiments upon 'The Belvidere Apollo,' which she never saw, nor anything like it, and knows nothing about."

* * * * *

"I will excuse you from the poetry, if you will read what comes next."

"Thank you, my dear father, a thousand times. It would have made me as sick as a cup-full of warm water would do. You know I had rather take so much hot drops. But the next article is Miss Simpkins's very original tale, entitled 'The Injured One,' probably all about love and despair, and ladies so fair, and men who don't care, if the mask they can wear, and the girls must beware. Now ain't I literary? But to be a heroine also, I will muster my resolution, and commence the story:

"'Madeline and Emerilla were the only daughters of Mr. Beaufort, of H., New Hampshire.'

"Now, pa, I cant't go any farther: I would as lieve travel through the deserts of Sahara, or run the gauntlet among the Seminoles, as to wade through this sloshy story. Miss Simpkins always has such names to her heroines: and they would do very well if they were placed any where but in the unromantic towns of our granite State."

* * * * *

"What comes next?"

"The editorial, pa, and the caption is, 'Our Representatives.' I had ten times rather read about the antediluvians, and I wish sometimes they might go and keep them company.

"And now for the items: Our new bell got cracked, in its winding way to this 'ere town; and the meeting-house at the West Parish has been fired by an incendiary; and the old elm near the Central House, has been blown down; and Widow Frye has had a yoke of oxen struck by lightning; and old Col. Morton fell down dead, in a fit of apoplexy; and the bridge over the Branch needs repairing; and 'a friend of good order' wishes that our young men would not stand gaping around the meeting-house doors, before or after service; and 'a friend of equal rights' wishes that people might sell and drink as much rum as they please without interference, &c. &c.; and all these things we knew before, as well as we did our A B Cs. Next are the cards: The *ladies* have voted their thanks to Mr. K., for his lecture upon phrenology, the matrimonial part, I presume, included: and the Anti-Slavery Society is to have a fair, at which will be sold all sorts of abolition things, such as anti-slavery paper, wafers, and all such important articles. I declare I will make a nigger doll for it. And Mr. P. of Boston, is to deliver a lecture upon temperance; and the trustees of the Academy have chosen Mr. Dalton for the Preceptor, and here is his long advertisement; and the Overseers of the Poor are ready to receive proposals for a new alms-house; and all these things, pa, which have been the town talk this long time. But here is something new. Our minister, dear Mr. Olden, has been very seriously injured by an accident upon the Boston and Salem railroad. The news must be very recent, for we have not heard of it; and it is crowded into very fine type. Oh, how sorry I am for him!"

"Well, Lina, or Miss Angeline, *there is something* of sufficient importance to repay you for the trouble of reading it, and I am very glad that you have done so, for I will start upon my intended journey to Boston to-day, and can assist him to return home. Any thing else?"

"Oh, yes, pa! a long list of those who have taken advantage of the Bankrupt Act, and the Deaths and Marriages; but all mentioned here, with whose names we were familiar, have been subjects for table-talk these several days."

"Well, is there no foreign news?"

"Yes, pa; Queen Victoria has given another ball at Buckingham Palace; and Prince Albert has accepted of a very fine blood-hound, from Major Sharp, of Houston; and Sir Howard Douglas has been made a Civil Grand Cross of the Bath, &c. &c. Are not these fine things to fill up our republican papers with?"

"Well, my daughter, look at the doings in Congress: that will suit *you.*"

"You know better, pa. They do nothing there but scold, and strike, and grumble — then pocket their money, and go home. See, here it begins: 'The proceedings of the house can hardly be said to have been *important.* An instructive and delightful *scene* took place between Mr. Wise, of Virginia, and Mr. Stanley, of South Carolina.' Yes, pa, that's the way they spend their time. In this *act* of the farce, or tragedy, one called t'other a *bull-dog*, and t'other called one a *coward.* Do you wish to hear any more?"

"You are somewhat out of humor, my child; but are there no new notices?"

"Yes, here is an 'Assessors' Notice,' and an 'Assignee's Notice,' and a 'Contractor's Notice,' but you do not care any thing about them. And here is an 'Auction Notice.'"

"What auction? Read it, my love."

"Why, the late old Mr. Gardner's farm-house, and all his furniture are to be sold at auction. And here is a notice of a meeting of the Directors of the Pentucket Bank, to be held this very afternoon."

"I am very glad to have learned of it, for I must be there. Is that all?"

"All? — no, indeed! Here are some long articles, full of *Whereases*, and *Resolved's*, and *Be it enacted's;* but I know you will excuse me from reading them. And now for the advertisements: Here is a fine new lot of *Chenie-de-Laines*, 'just received' at Grosvenor's — oh, pa! do let me have a new dress, won't you?"

"No, I can't — at least I don't see how I can. But if you will promise to read my papers through patiently for the future, and will prepare my valise for my journey to Boston, I will see what I may do. Meantime

I must be off to the directors' meeting. And now let me remind you that two items, at least, in this paper, have been of much importance to me ; and one, it seems, somewhat interesting to you. So no more fretting about the Chronicle, if you want a *new gown*."

Mr. Wheeler left the room, and Angeline seated herself at the work-table, to repair his vest. She was sorry she had fretted so about the Chronicle ; but she did wish her father would take the "Ladies' Companion," or something else in its stead.

While seated there, her little brother came running into the room, all out of breath, and but just able to gasp out, "Oh, Lina ! there is a man at the Central House, who has just stopped in the stage, and he is going right on to Kentucky, and straight through the town where Alfred lives, for I heard him say so ; and I asked him if he would carry anything for us, and he said, 'Yes, willingly.' So I ran home as fast as I could come, to tell you to write a note, or do up a paper, or something, because he will be so sure to get it, and right from us, too, as fast as it can go. Now do be quick, or the stage will start off."

"Oh, dear me !" exclaimed Angeline, how I do wish we had a New York Mirror, or a Philadelphia Courier, or a Boston Gazette, or any thing but this stupid Chronicle ! Do look, Jimmy ! is there nothing in this pile of papers ?"

"No, nothing that will do, so fold up the Chronicle, quick, for the stage is starting."

Angeline, who had spent a few moments in looking for another paper, now had barely time to scrawl the short word "Lina" on the paper, wrap it in an envelope, and direct it. Jimmy snatched it as soon as it was ready, and ran out "*full tilt*," in knightly phrase, or as he afterwards said, "*lickity split*."

The stage was coming on at full speed, and he wished to stop it. Many a time had he stood by the road-side, with his school companions, and, waving his cap, and stretching out his neck, halloed, "Hurrah for Jackson !" and he feared that, like the boy in the fable, who called "Wolves ! wolves !" if he now shouted to them from the road-side, they would not heed him. So he ran into the middle of the road, threw up his arms, and stood still. The driver barely reined in his horses within a few feet of the daring boy.

"Where is the man who is going straight ahead to Kentucky ?"

"Here, my lad," replied a voice, as a head popped out of the window, to see what was the matter.

"Well, here is a paper which I wish you to carry to my brother ; and if you stop long enough where he is, you must go and see him, and tell him you saw me too."

"Well done, my lad! you are a keen one. I'll do your bidding, but don't you never run under stage-horses again."

He took the packet, while the driver cracked his whip; and the horses started as the little boy leaped upon the bank, shouting, "Hurrah for Yankee Land and old Kentucky!"

In a rude log hut of Western Kentucky was seated an animated and intelligent-looking young man. A bright moon was silvering the forest-tops, which were almost the only prospect from his window; but in that beauteous light the rough clearing around seemed changed to fairy land; and even his rude domicil partook of the transient renovation. His lone walls, his creviced roof, and ragged floor, were transformed beneath that silvery veil; and truly did it look as though it might well be the abode of peaceful happiness.

The young man soliloquising on the incitement to write poetry, and then, reflecting on his lone state, his heart turns towards home, and he proceeds: —

"How I wish I could see them all! There is not a codger in my native town, not a crusty fusty old bachelor, an envious tattling old maid, not a flirt, sot, pauper, idiot, or sainted hypocrite, but I could welcome with an embrace. But if I could only see my father, or Jimmy, or Lina, dear girl! how much better I should feel! It would make me ten years younger, to have a chat with Lina; and, to tell the truth, I should like to see any woman just to see how it would seem. I'd go a quarter of a mile, now, to look at a row of aprons hung out to dry. But there! it's of no use to talk.

"An evening like this is such an one as might entice me to my mother's grave, were I at home. Oh! if she were but alive, if I could only know that she was still somewhere on the wide earth, to think and pray for me, I might be better as well as happier. Methinks it must be a blessed thing to be a mother, if all sons cherish that parent's memory as I have mine — and they do. It cheers and sustains the exile in a stranger's land; it invigorates him in trial, and lights him through adversity; it warns the felon, and haunts and harrows the convict; it strengthens the captive, and exhilarates the homeward-bound. Truly must it be a blessed thing to be a mother!"

He stopped — for in the moonlight was distinctly seen the figure of a horseman, emerging from the public road, and galloping across the clearing. He turned towards the office of the young surveyor, and in a few moments the carrier had related the incident by which he obtained the paper, and placed "The Village Chronicle" in Alfred's hand.

He struck a light, tore off the wrapper, and the only written word which met his eye was "Lina." "Dear name!" said he, "I could almost kiss it, especially as there is none to see me. She must have been in a prodigious hurry; and how funny that little rascal, Jimmy, must have looked! Well, 'when he next doth run a race, may I be there to see.'"

He took the paper to read. It was a very late one—he had never before received one so near the date; and even that line of dates was now so pleasing. First was Miss Helena Carrol's poetry. "Dear girl!" said he, "what a beautiful writer she is! Really, this is poetry! this is something which carries us away from ourselves, and more closely connects us with the enduring, high, and beautiful. Methinks I see her now—more thin, pale, and ethereal in her appearance than when we were gay schoolmates; but I wonder that with all her treasures of heart and intellect, she is still Helena Carrol.

"And now here is Miss Simpkins's story of 'The Injured One'—beautiful, interesting and instructive, I am confident; and I will read it, every word; but she italicises too much; she throws too lavishly the bright robes of her prolific fancy upon the forms she conjures up from New England hills and vales. I wonder if she remembers now the time when she made me shake the old apple tree, near the pound, for her, and in jumping down I nearly broke my leg. Well, if I read her story, I will try that it do not break my heart.

"And here is an excellent editorial about 'Our Representatives,'—I will read it again, and now for the ITEMS."

These were all highly interesting to the *absentee*, and on each did he expatiate to himself. How different were his feelings from his sister's, as he read of the cracked bell, the burned meeting-house, the dead oxen, the apoplectic old colonel, the decayed bridge, the hints of the friends of "good order" and "equal rights!" Then there was a little scene suggested by every card. He wondered who had their heads examined at the phrenological lecture; and if the West Parish old farmers were now as stiffly opposed to the science. And how he would like to see Lina's chart, and to know if Jimmy had brains—he was sure he had legs, and a big heart, for a little boy; and he wondered what girls ran up to have their heads felt of in public; and what the man said about matrimony—an affair which in old times was thought to have more to do with the heart than with the head.

* * * * * *

Then they were to have a new alms-house; and forty more things were mentioned, of equal interest — not forgetting Mr. Olden's accident, for which "father would be so sorry." Then there were the Marriages and Deaths — each a subject of deep interest, as was also the list of Bankrupts. The foreign news was news to him, and Congress matters were not passed unheeded by.

Then he read with deep interest every "Assessor's Notice," also those of "Assignees," "Contractors," and "Auctioneers." There was not a single "Whereas," or "Resolved," but was most carefully perused; and every "Be it enacted" stared him in the face like an old familiar friend.

Then there were the advertisements; and Grosvenor's first attracted his attention, from its *big* letters. "'CHENIE-DE-LAINES!'" said he, "what in the name of common sense are they? Something for gals' gowns, *I guess*; and what will they next invent for a name."

But each advertisement told its little story. Some of the old "*pillars*" of the town were still in their accustomed places. The same signatures, places, and almost the same goods — nothing much changed but the dates. Another advertisement informed him of the dissolution of an old copartnership, and another showed the formation of a new one. Some old acquaintances had changed their location or business, and others were about to retire from it. Those whom he remembered as almost boys, were now just entering into active life, and those who should now be preparing for another world were still laying up treasures on earth. One, who had been a farmer, was now advertising himself as a *doctor*. A lawyer had changed into a miller, and old Capt. Prouty was postmaster. The former cobbler now kept the book-store, and the young major had turned printer. The old printer was endeavoring to collect his debts, for he said his devil had gone to Oregon, and he wished to go to the devil.

Not a single puff did Alfred omit; he noticed every new book, and swallowed every new nostrum. "Old Rags," "Buffalo Oil," "Bears' Grease," "Corn Plaster," "Lip Salve," "Accordions," "Feather Renovators," "Silk Dye-Houses," "Worm Lozenges," "Ready-made Clothing," "Ladies' Slips," "Misses' Ties," "Christmas Presents," "Sugarhouse Molasses," "Choice Butter," "Shell Combs," "New Music," "Healing Lotions," "Last Chance," "Hats and Caps," "Prime Cost," "Family Pills," "Ladies' Cuff Pins," "Summer Boots," "Vegetable Conserves," "Muffs and Boas," "Pease's Horehound Candy," "White Ash Coal," "Bullard's Oil-Soap," "Universal Panacea," "Tailoress Wanted," "Unrivalled Elixir," "Excellent Vanilla," "Taylor's Spool Cotton," "Rooms to Let," "Chairs and Tables," "Pleasant House," "Particular Notice," "Family Groceries," "A Removal," "Anti-Dyspeptic Bitters," &c., &c., down to "One Cent Reward — Ran away from the Subscriber," &c. Yes, he read them all, and all with much interest; but one with a deeper feeling than was awakened by the others. It was the notice of the sale of the late Mr. Gardner's house, farm, &c.

"And so," said Alfred, "Cynthia Gardner is now free. She used to love me dearly — at least she said so in everything but words; but the old man said she should never marry a harum-scarum scape-grace, like me. Well! it 's no great matter if I did sow all my wild oats then, for there is too little cleared land to do much at it here. The old gentleman is dead, and I'll forgive him; but I will write this very night to Cynthia, and ask her to —

——'come, and with me share
Whate'er my hut bestows;
My corn-stalk bed, my frugal fare,
My labor and repose.'"

LUCINDA.

Now which of our factory operatives could write such a description and story as that? Yet why should they not? Are they inferior in natural intelligence, or in mental abilities, to their sisters — almost their *own* sisters — of New England? Or, if not, is there anything essential to their state or employment, that should hinder them from, at all events, *aspiring* to an improvement in their condition? Certainly there is not. Then, let me tell you, my children, that if *you* are willing to do all *you* are able to do, for raising yourselves, in dependence on the Divine blessing, to a higher moral, social, and intellectual condition — there are those amongst you who will be ready to assist you in the worthy enterprise, and delighted to help you.

Not to anticipate too far, the suggestions I shall have to offer to you in a subsequent Lecture, in the form of a *plan* for the object commended to you, it may suffice at present just to say, that if but our factory-operatives and factory-masters, and the Christian ministers, and the Christian people of our general population, would cordially take up this matter on the principles of their Christian profession, (as I fervently trust they will be induced to do,) benefit, real and substantial, must ensue. And though our factory-girls might not be raised to the literary rank of writers in a "BRADFORD OFFERING," they would be placed, as a class, in a condition far more honorable to themselves than that which at present they enjoy, and rendered vastly more useful to society. Yes, and in place of the sad and lamentable profligacy, which so awfully, I fear, prevails, we should gain for them a higher tone of moral and religious principle, and see, by the blessing of God, numbers elevated to the true dignity of our race, — to the character which felicitates and adorns — to the state which elevates woman to her proper condition here, and provides for her an ennobling, immortalizing destiny hereafter.

That there are available, and, under the Divine blessing, adequate means capable of being employed for advancing these great and Christian objects, I hope in the succeeding Lecture to be able to show; and I pray God that the minds of you all who might be able to help us may be inclined to do so with a free, hearty good will, for the benefitting of our factory operatives, and for the promotion thereby of the Divine glory.

CHAPTER III.

THE FEMALE OPERATIVES, OR "FACTORY-GIRLS" OF LOWELL.

Already have I described to you the factories and female operatives at Lowell, as they appeared to me on my visit there. But from the enquiries made on various occasions, and from information derived from other sources, I was enabled to make a much better acquaintance with Lowell, and its factory system, than a brief visit, as a stranger, could possibly afford.

One leading design of this Lecture being to place before you a particular account of the female operatives, or, as they are usually designated, "The Factory Girls of Lowell," I shall now avail myself of the different sources of information open to me for this object, in the earnest hope that something of what shall be found worthy of peculiar commendation in these American factories may be made profitable for our imitation.

Several distinct points, of interesting consideration concerning the *factory-girls of Lowell*, will successively engage our attention, such as, — Their previous condition; their object in joining the factory-operatives; their boarding-house system, as described and illustrated by themselves; the period of labor and manner of working in the factories; their recreations and leisure employments; and their general character and moral condition.

Sect. 1.—*As to the previous condition of the Factory Girls of Lowell.*

Who or what are they? and where do they come from? are enquiries of some interest concerning a class of operatives, whose educational acquirements and moral character have excited so much attention with strangers, and especially with those who feel an interest in the improvement of the condition of our working population.

The Lowell female operatives are not, as with us they commonly are, young persons who have been brought up as children in the factories—for there are very few children employed at Lowell. Nor are they generally the children of persons engaged in the local factories, or otherwise as laborers. Nor are they prevalently, the children of the poorest and humblest classes, as they usually are with us. Nor are they, except in a minor proportion, the children of persons residing in the town or in the neighborhood.

For in respect of the source from whence the female part of the hands are supplied, Lowell presents a remarkable peculiarity of which we have no example in this country. With us, the families who supply the hands for our factories, parents as well as children, are, usually, either resident amongst us, or in the neighboring villages. And when in prosperous times, additional hands are required, the demand, on account of the wages, attracts the requisite supply, and a stream of new population flows into the town and neighborhood; and these, by the building of additional cottages, and of whole streets of cottages, (which their necessities call for,) being promptly provided with residences, usually settle in the place.

In Lowell, however, the case is far otherwise. The

resident families, no doubt, supply a good many of the hands; but by much the larger proportion of the young women is supplied from the country, near and remote, around. Many of these, coming from a distance of fifty to a hundred miles or more, and some out of other States, leave their parental home and natural protectors, and place themselves confidingly under the arrangements made by the different manufacturing corporations for their reception.

And to these factory operatives from a distance, a preference, I believe, is given by the managers, as affording a safe supply under the contingency of slack times,—the occasions on which our manufacturing population is liable greatly to suffer. For these give safety, as by the escape-steam, to the moral pressure in the factories when the quantity of duty is not sufficient to work off the supply of physical force, by being able to return to their homes, where they can be sufficiently provided for as before, without any suffering to the community, and, generally, without great inconvenience to themselves.

The class to which they usually belong is that of *farmers*, many being the daughters of able and independent yeomen. For it should be remembered, that the American farmer is generally, and all but universally, the proprietor of the land which he cultivates, an *independency of condition* arising, not from anything in their republican constitution, but from the immense extent of the national territory, and the consequent cheapness of the unoccupied lands. These farmers, however, able and well off as many of them are, do not (as to a large number at least) consider it either discreditable, or unsafe, for their daughters to leave their domestic fireside to become factory operatives.

Independent as the circumstances of some of the parents are, and respectable and well-educated as their daughters may be, they are neither above working themselves, nor letting their daughters work. To labor, indeed, with one's hands is not held to be, nor should it be, a demeaning employment even for those who might be able to live without it; for a country like America needs all the physical powers of its population, and could employ, perhaps, a hundred-fold more. Hence, were every one, however wealthy, so to labor, no injury, such as would happen with us, would be done to the poor; because there is room and occupation for all.

But for the country farmers and for their families, there are special advantages to be gained by this employment of their young females, which I now proceed to notice, by speaking in the next place,—

SECT. 2.—*Of their object in joining the Factory Operatives.*

With the farmer in America, the earth brings forth grain, with all manner of fruits and herbs, abundantly. Most abundantly doth it "give seed to the sower and bread to the eater." But one essential for domestic comfort it does not so readily yield, that is *money:* for the produce of multitudes of farms, plentiful as it is for domestic uses, is but little available for being turned into money. Most of the produce of the far country farm, indeed, bears but a low price at the nearest market; and such market is, not unfrequently, at a most inconvenient distance. Hence an employment, like that afforded at Lowell, where liberal wages are paid, and money can be easily saved, becomes economically important, and, for the personal benefit, very attractive.

Whilst the obtaining of *spare* money, in sure and progressive accumulation, may be stated as the general object by which the country girls are induced to become factory operatives, the desire for the gain they can make, is stimulated by a variety of motives, some, no doubt, by motives of personal vanity, some of self-aggrandizement, and some, which I delight to name, by considerations which do honor to the best feelings of humanity.

Thus have I heard or read of cases—to begin with the lowest motive—where a country girl, seeing a village associate return from Lowell clothed in silks, and otherwise gaily adorned, has, from the mere feeling of envy, been urged to seek the like employment, that she might be indulged in similar vanity. Other cases are well known, and these perhaps of the most extensive occurrence, where the young people, having no means of acquiring any personal property at home, nor their parents of giving them any, go to Lowell in order to procure a little fund, out of their earnings, for their benefit in marrying, and in their settlement in life. Others, again, become factory operatives, of which we find many examples, for much higher and more dignified objects,—such as for the relief of an anxious father from a burden of debts, or for his release from the pressure of a mortgage on his property,—the claims for the high interest on which had caused him, perhaps, to struggle very hardly. And cases are well known, where an affectionate sister has left her parental roof to become a factory girl, that she might obtain the means, by her personal toils, of benefitting her younger sisters or brothers, in education, or of obtaining a college education, as we believe to have been the fact, for a beloved brother!

These are cases, stated to have occurred, and which I can well believe, honorable alike to humanity, and to the high-minded, self-denying girls, who could be actuated by such noble motives.

In the instance of *widows*, left with young children, Lowell, as with ourselves, affords a mean, by the labor of their hands, of providing for their subsistence; and in the class of married women (a class amongst our factory workers, I should rejoice to see lessened, and, as far as possible, abolished), Lowell has also its factory operatives; but the principal class, and by far the most numerous, I believe, is that which I have more particularly described, as the young daughters of the provincial farmers.

The earnings of the factory operatives, and their possible weekly or annual savings, I have already stated; but the actual results of their savings, as indicated by the deposits in the "Lowell Institution for Savings," should not be omitted. The number of depositors in this Institution, in July, 1840, (?) was 1976, and the amount of their deposits above three hundred thousand dollars. Of these depositors, 978 were *factory girls*, and their proportion of the deposits about *one hundred thousand dollars*, or above £20,000! And it is a common thing for one of these girls to have five hundred dollars (a hundred guineas, nearly) in deposit; and the amount only stops here, because no interest being paid for higher sums, some other investment is sought for the extra savings.*

In regard to *children*, I have already noticed, that but few, proportionally, are employed in these factories. The

* Dr. Bartlett's Vindication, p. 22.

humane and paternal provision for securing time for the education of children, may, possibly, act as a check to their more extensive employment. For "by a statute of the Commonwealth it is provided, that no person *under the age of fifteen years*, shall be suffered to work more than nine months of any year in a manufacturing establishment, the remaining three months to be passed at school." And to secure the fulfilment of this benevolent rule, "every child is obliged to bring, and she does bring it," observes Dr. Bartlett, from whom I quote, "before she can be admitted into the mill, a certificate from a school teacher, signed by him or her, and sworn to before a justice of the peace, that the terms of the statute have, in her case, been truly and faithfully complied with." And Dr. Bartlett further avers, that the law, thus appointed, is enforced with unflinching strictness; so that even in hard times, or in times of scarcity of hands, no plea, in any instance, that he ever heard of, was admitted for the wresting of the law from its important design.

The *average age* of the operatives employed in these factories, differs greatly from what occurs with us. In Dr. Ure's Philosophy of Manufactures, it is, I believe, stated, that of 8289 males, and 8946 females employed in 43 cotton-mills in Manchester, 2338 males and 1589 females were betwixt the ages of nine and fourteen years; whilst the average age of the operatives in the mills at Preston and its vicinity, according to a report of Mr. Harrison, Inspecting Surgeon, was, in 1834, fourteen years.

But in Lowell, though the general fact, as to the average age has not been determined, yet within particular limits it has. And the result was this: that of near 2000 girls, in whose case this point was ascertained, the average age was about twenty-three years. In one establishment, employing 657 young women, the exact mean age was found to be twenty-two years and a quarter; and the average time they had been working in factories three years and three quarters. In another establishment, a single factory only, employing 203, the mean age was twenty-three years, nearly; and the average time during which they had worked in factories about four and a quarter years.

(3.) Next, let me tell you more particularly, about *their boarding-house system, as described and illustrated by themselves.*

The mention I have already made of the Lowell boarding-houses, related, chiefly, to what my own eyes saw in a hasty visit. But it may be interesting to our Bradford factory operatives to know something more about what goes on therein.

The reason for the introduction of this system, so different from ours, you will have anticipated. Of the 7000 females, or thereabout, employed in the Lowell factories, a large proportion have neither parents nor relatives resident, consequently, no home there. Provision for their accommodation, therefore, is made, not as with us, when such is required, by the chance supply of ordinary and inferior lodging houses; but by the careful pre-arrangement, chiefly, of the respective corporations. Houses of the neat and commodious class, which I have already described, are erected, and placed under the superintendence of respectable females, as the matrons. In order to their affording cheapness of living for the female operatives, the tenants pay but a small rental for the houses, — a charge which, sometimes, when they are slack of boarders, is partially remitted. They provide generally, I believe, a fair supply of substantial food, and more than is perhaps needful of variety, — they being reimbursed by the very moderate payments, before mentioned, of the inmates.

Of the plan and management of these boarding-houses, I shall be able to give you a good idea, by reading to you a part of a very characteristic letter published in the Lowell Offering, in the number for May last, 1844, — descriptive of the feelings and observations of a young woman from the country on her arrival, as a stranger, at Lowell, and on her being received into one of the lodging houses wherein she found at first, neither friend nor acquaintance. This letter is one of a series of three, which were published when I was in America, and profess to be written by "Susan," from Lowell, to her friend "Mary," in the country. The account she gives is so very natural, clever, and descriptive, that I fancy you will listen to it, as if I were reading to you her own written letter. Susan thus begins: —

Lowell, March —, ——.

Dear Mary: — When I left home I told you that I would write in a week, and let you have my first impressions of Lowell. I will keep my promise; though, if I should defer my letter a while longer, I think I could make it much more interesting. But you know I promised to be very minute, and there is always sufficient minutiæ to fill up a letter.

I arrived here safe and sound, after being well jolted over the rocks and hills of New Hampshire; and when (it was then evening) a gentleman in the stage first pointed out Lowell to me, with its lights twinkling through the gloom, I could think of nothing but Passampscot swamp, when brilliantly illuminated by "lightning-bugs," (Fire-flies.) You, I know, will excuse all my "up-country" phrases, for I have not yet got the rust off; and to you and all my old-fashioned friends, I shall always be *rusty*. My egotism I will not apologize for — it is what you request.

To return to my adventures — for it all appears very romantic to me. The driver carried me to the "corporation," as it is called; and which, so far as I now can describe it, is a number of short parallel streets with high brick *blocks* on either side. There are some blocks with blinds to them, and some are destitute. Some of the doors have bells, others have not. Contiguous to these *boarding-houses* are the *mills*, of which I will tell you more by and by.

I told the driver to carry me to No. —, and there he left me; where there was not a soul that I knew, if cousin Sarah was gone. I inquired of an Irish girl who came to the door, if Sarah G. Pollard boarded there. She said that she had gone to Manchester, to work with an overseer who was an old acquaintance. The girl did not invite me in, and there I stood like a "statter," as Aunt Hitty says. I did not feel disposed to make inquiries of the girl, I was so unaccustomed to her brogue. Just then — that is, just as my heart was sinking ten fathoms below zero — a pleasant-looking woman came into the entry; and in a very motherly way, invited me into her own room; took off my things, ordered away my trunk and bandbox, brought camphor for my head, for it ached with my ride, and told me all about cousin Sarah. She said that I had better not think of following her to Manchester, and promised to do all for me that she could. This was Mrs. C., "the boarding-woman" — a widow, with several children, whom she keeps at school, and maintains well, by her own industry and good management.

I had expected coldness, or at least entire indifference, in this city, and the cordiality of the good landlady filled my heart with gratitude. I have since inquired if she were not unusually kind, but, though she is a very good woman, the girls here say that she is not more so to me than to any other new boarder; and that the boarding-women are always "dreadful good" to a new boarder. Every girl, let her be ever so rusty, or rather rustic, fills one of the many niches prepared here for so many, and some, you know, are like nest eggs, and bring many more. But we will not be so uncharitable as to suppose there is nothing but policy in all this, for there is surely something to excite a woman's sympathies in the sight, which is not uncommon here, of a lonely, friendless, helpless stranger.

You can hardly think how my heart beat when I heard the bells ring for the girls to come to supper; and then the doors began to slam, and then Mrs. C. took me into the dining-room, where there were three common-sized dining tables, and she seated me at one of them, and then the girls thickened around me, until I was almost dizzy.

At the table where I sat they were very still, for the presence of a stranger is usually "a damper" upon them. But there was quite noise enough at the other tables, and what was wanted in wit was made up in merriment. After a while one or two of my boon companions "opened their mouths and spoke," and I have already found that those who make themselves most conspicuous in the presence of strangers, and would soonest attract their attention, are those who do themselves, and those with whom they are connected, the least credit.

I remember that I must be very minute — so I will inform you that we had tea, flapjacks, and plum-cake for supper. There was also bread, butter, and crackers, upon the table ; but I saw no one touch them.

After supper the tables were cleared in a trice. Some of the girls came in with their sewing, some went to their own rooms, and some went "out upon the street," — that is, they went to some meeting, or evening school, or they were shopping, or visiting upon some other corporation, all of which is called "going upon the street," in factory parlance.

I retained my seat with the girls in the great keeping-room, for Mrs. C. had company in her own sanctum, and I did not know where else to go. * * * * * *

At ten o'clock Mrs. C. came in, and told us that it was time for us all to go to bed. Some begged for time to "read this story out;" others just for a "few minutes to finish this seam." She refused them good-naturedly, but those were most cunning who wanted to warm their feet, and detained her by telling queer stories, of what they had seen and heard upon "the street," — and she unconsciously gave them the few minutes she had at first refused.

I was shown up three flights of stairs, into what is called "the long attic," — where they put all poor stranger girls, — the most objectionable places being always left for new comers. There were three beds in it, only two of which were occupied, for this is always the room for vacancies. My baggage had already been carried up by "the boys," as the boarders call Mrs. C.'s sons ; and I looked wofully at the strange girl who was to be my "*chum*." She took no notice of me, and went to sleep as composedly as if I had been still among the White Mountains ; but the two girls in the further bed kept whispering together something about "the old man." I was very nervous, and almost wished "the old boy" had them both ; but, when the house was still, a strange fear came over me, such as is created in children by telling them about *the old man.*

I heard the bells strike the midnight hour long before I went to sleep, and then I dreamed about "the old man."

As soon as the day broke I was awakened by one of the girls jumping out of bed and beginning to crow. That awakened the others, and they bestirred themselves. One sung

> Morning bells I hate to hear.
> Ringing dolefully, loud, and drear, &c.

Then the other struck up, with a loud voice,

> Now isn't it a pity,
> Such a pretty girl as I,
> Should be sent to the factory
> To pine away and die.

I dressed myself and followed them down stairs, where I found my place at the table, and our early breakfast was all ready for us. It consisted of hot cakes and coffee; there was also "hash" upon the table for those who wanted it.

After the girls had gone to work, Susan informs us, that she applied to Mrs. C. for directions, as to what she should do, to get employment; and that Mrs. C. went out to inquire about a place for her. "No overseer, however," as Susan continues in her letter, "would take me, even upon her recommendation, until they had seen me themselves. One promised, however, to give me work if he likes the looks of me, and she considers this place as already engaged, for, she says, she knows he will like me when he sees me."

The overseer at length came, and approved of the applicant, and engaged her as an operative; but could not receive her till the next week.

Meanwhile she had an opportunity of seeing the place, —"the lions and the lionesses, and of getting an idea of her location before going into the mill." The result of her explorations she gives in the same letter — noticing or describing the principal streets — "stores filled with beautiful goods," — public buildings — the Merrimack House

— the railroad depôt, — alluding, also, to her first impressions on seeing the arrival of a train of " cars," and concerning the first sight of a city.

I give the next part, in her own words, because, common-place as the subject is, the descriptions are characteristic.

When I went out with Mrs. C. she made me put on one of her girls' bonnets, because mine did not turn up behind, and out at the ears, and she said it was O. S., instead of O. K. Well, as I walked along, and saw all the beautifully dressed ladies, I thought, within myself, that, with bonnets and dresses of an old style, they too would not be passable.

You must know that they dress very much here, — at least, it so appears to us, who have just come off of the hills, and been accustomed to put on our woollen gowns in the morning, and our better woollen gowns in the "arternoon." Here they wear velvets, and furs, and plumes, and bugles, and *all*. I should wish to know a great deal to be dressed so, for I should think there was a great deal to be expected of one who made such pretensions.

I told Mrs. C. that the city ladies were not so pale as I expected. She said that many of them were painted, and that *rouge* was becoming more fashionable every year. She says that even some of the factory-girls use it, and pointed out several highly dressed girls whose cheeks were truly of "a carmine tint."

The last remark I have read to you, not as a subject for commendation, nor as a practice at all likely to be followed in a manufacturing town in England, but because it is descriptive. It goes to the foibles of the young women — that is, as to *some* of them — and shows how amid so much that is exemplary and worthy of imitation, human nature and female vanity have their influence at Lowell, the same as elsewhere; and, in a society so peculiar, these influences are shown, not only in an excessive love of dress, but even in the producing of a fictitious appearance of healthful ruddiness to hide the supposed defect of too great paleness of the face.

SECT. 4.—*The period of labor, and the manner of working the factories,*—may require a brief description before we proceed to speak more of the factory operatives.

As to the manner of working their factories (except their water power), there is, I believe, nothing essentially different from ours. In the arrangement of the machinery, and the appropriation of particular rooms or floors for each department of work, I did not perceive anything particular.

In Susan's second letter to Mary, published in the Lowell Offering for June 1844, a very explicit and characteristic description is given of the various kinds of work attended to by the girls in the cotton factories. The hardest work—or at least that which by a voluntary attendance on more than a pair of looms requires the most constant and vigilant attention, is that of the weavers. A pair of looms is the ordinary allotment; but if a weaver can superintend three or four looms, for which she obtains an encouraging addition of pay, the work becomes very trying.

The plan of working the mills, as to hours for working and meals, varies with the seasons of the year. In the letters of Susan, just referred to, this account is given of the hours of labor:—

"We go in," she says, speaking of the summer season, "at five o'clock; at seven we come out to breakfast; at half-past seven we return to our work, and stay until half-past twelve. At one, or a quarter-past one, four months in the year, we return to our work, and stay until seven at night."

In the early part of the year, she mentions this different arrangement:—

"When I first came here, which was the last of February, the girls ate their breakfast before they went to their work. The first of March they came out at the present breakfast hour (seven); and the 20th of March they ceased to light up the rooms, and came out betwixt six and seven o'clock......Then the evening is all our own, which is more than some laboring girls can say, who think that nothing is more tedious than a factory life."

It thus appears, that — with the exception of particular periods betwixt the lighting up of the mills, and the working entirely by daylight — the total number of hours of running the mills, from starting to stopping, is fourteen; and as only two intervals appear to be allowed for meals, amounting together to an hour, or an hour and a quarter, the time during which they actually work, must, in the summer, be nearly thirteen hours a day! The *average* period of *actual labor* throughout the year, appears, however (from a variety of statements that I have met with on this point) to be about twelve hours.

The effect of this extent of attention, and long time of standing, on the physical condition of the factory operatives, is thus described by Susan: —

It makes my feet ache and swell to stand so much, but I suppose I shall get accustomed to that too. The girls generally wear old shoes about their work, and you know nothing is easier; but they almost all say that when they have worked here a year or two they have to procure shoes of a size or two larger than before they came. The right hand, which is the one used in stopping and starting the loom, becomes larger than the left; but in other respects the factory is not detrimental to a young girl's appearance. Here, they look delicate, but not sickly; they laugh at those who are much exposed and get pretty brown; but I, for one, had rather be brown than pure white. I never saw so many pretty looking girls as there are here.

In the same letter (the second) Susan describes, with her usual neat and characteristic simplicity, her own commencement and progress.

Dear Mary: — In my last I told you I would write again, and say more of my life here; and this I will now attempt to do.

I went into the mill to work a few days after I wrote to you. It looked very pleasant at first, the rooms were so light, spacious, and clean, the girls so pretty and neatly dressed, and the machinery so brightly polished or nicely painted. The plants in the windows, or on the overseer's bench or desk, gave a pleasant aspect to things. You will wish to know what work I am doing: —

After describing the various kinds of work undertaken by the female operatives in the different departments of a cotton-factory, and their respective denominations, such as, carders, spinners, doffers, dressers, drawers-in, and weavers, she thus proceeds: —

The dressing-rooms are very neat, and the frames move with a gentle undulating motion which is really graceful. But these rooms are kept very warm, and are disagreeably scented with the "sizing," or starch, which stiffens the "beams," or unwoven webs. There are many plants in these rooms, and it is really a good green-house for them. The dressers are generally quite tall girls, and must have pretty tall minds too, as their work requires much care and attention.

I could have had work in the dressing-room, but chose to be a weaver; and I will tell you why. I disliked the closer air of the dressing-room, though I might have become accustomed to that. I could not learn to dress so quickly as I could to weave, nor have work of my own so soon, and should have had to stay with Mrs. C. two or three weeks before I could go in at all, and I did not like to be "lying upon my oars" so long. And, more than this, when I get well learned I can have extra work, and make double wages, which you know is quite an inducement with some.

Well, I went into the mill, and was put to learn with a very patient girl — a clever old maid. I should be willing to be one myself if I could be as good as she is. You cannot think how odd everything seemed to me. I wanted to laugh at everything, but did not know what to make sport of first. They set me to threading shuttles, and tying weaver's knots, and such things, and now I have improved so that I can take care of one loom. I could take care of two if I only had eyes in the back part of my head, but I have not got used to "looking two ways of a Sunday" yet.

At first the hours seemed very long, but I was so interested in learning that I endured it very well; and when I went out at night the sound of the mill was in my ears, as of crickets, frogs, and jews-harps, all mingled together in strange discord. After that it seemed as though cotton-wool was in my ears, but now I do not mind it at all. You know that people learn to sleep with the thunder of Niagara in their ears, and a cotton-mill is no worse, though you wonder that we do not have to hold our breath in such a noise.

Certain inquiries of Susan's friend and correspondent, in the country, gave occasion for the introduction of a variety of other topics in this second letter. I select, either on account of the information conveyed therein, or of the recommendations before stated, the following: —

You ask if the girls are contented here: I ask you, if you know of *any one* who is perfectly contented. Do you remember the old story of the philosopher, who offered a field to the person who was contented with his lot; and, when one claimed it, he asked him why, if he was so perfectly satisfied, he wanted his field. The girls here are not contented; and there is no disadvantage in their situation which they do not perceive as quickly, and lament as loudly, as the sternest opponents of the factory system do. They would scorn to say they were contented, if asked the question; for it would compromise their Yankee spirit — their pride, penetration, independence, and love of "freedom and equality," to say that they were *contented* with such a life as this. Yet, withal, they are cheerful. I never saw a happier set of beings. They appear blithe in the mill, and out of it. If you see one of them with a very long face, you may be sure that it is because she has heard some bad news from home, or because her beau has vexed her. But, if it is a Lowell trouble, it is because she has failed in getting off as many "sets" or "pieces" as she intended to have done; or because she had a sad "break out," or "break down," in her work, or something of that sort.

You ask if the work is not disagreeable. Not when one is accustomed to it. It tried my patience sadly at first, and does now when it does not run well; but, in general, I like it very much. It is easy to do, and does not require very violent exertion, as much of our farm-work does.

You also ask how I get along with the girls here. Very well indeed; only we came near having a little flurry once. You know I told you I lodged in the "long attic." Well, a little while ago, there was a place vacated in a pleasant lower chamber. Mrs. C. said that it was my "chum's" turn to go down stairs to lodge, unless she would waive her claim in favor of me. You must know that here they get up in the world, by getting down, which is what the boys in our debating society used to call a paradox. Clara, that is the girl's name, was not at all disposed to give up her rights, but maintained them staunchly. I had nothing to do about it; the girls in the lower room liked me, and disliked Clara, and were determined that it should not be at all pleasant weather there if she did come. Mrs. C. was in a dilemma. Clara's turn came first. The other two girls in the chamber were sisters, and would not separate, so they were out of the question. I wanted to go, and knew Clara would not be happy with them. But I thought what was my duty to do. She was not happy now, and would not be if deprived of her privilege. She had looked black at me for several days, and slept with her face to the wall as many nights. I went up to her and said, "Clara, take your things down into the lower chamber, and tell the girls that *I will not come*. It is your turn now, and mine will come in good time."

Clara was mollified in an instant. "No," said she, "I will not go now. They do not wish me to come, and I had rather stay here." After this, we had quite a contest — I trying to persuade Clara to go, and she trying to persuade me, and I "*got beat*." So now I have a pleasanter room, and am quite a favorite with all the girls. They have given me some pretty plants, and they go out with me whenever I wish it, so that I feel quite happy.

You think we must live very nice here to have plum-cake, &c. The plum-cake, and crackers, and such things as the bakers bring upon the corporations, are not so nice as we have in the country, and I presume are much cheaper. I seldom eat anything that is not cooked in the family. I should not like to tell you the stories they circulate here about the bakers, unless I *knew* that they were true. Their brown bread is the best thing that I have tasted of their baking.

You see that I have been quite *minute* in this letter, though I hardly liked your showing the former to old Deacon Gale, and 'Squire Smith, and those old men. It makes me feel afraid to write you all I should like to, when I think so many eyes are to pore over my humble sheet. But if their motives are good, and they can excuse all defects, why I will not forbid.

'Squire Smith wishes to know what sort of men our superintendents are. I know very well what he thinks of them, and what their reputation is up our way. I am not personally acquainted with any of them; but from what I hear, I have a good opinion of them. I suppose they are not faultless, neither are those whom they superintend; but they are not the overbearing tyrants which many suppose them to be. The abuse of them which I hear, is so very low that I think it must be unjust and untrue; and I do frequently hear them spoken of as *men* — whole-hearted, full-souled men. Tell 'Squire Smith they are not what he would be in their places — that they treat their operatives better than he does his "hired girls," and associate with them on terms of as much equality. But I will tell you who are almost universally unpopular: the "runners," as they are called, or counting-room boys. I suppose they are little whipper-snappers who will grow better as they grow older.

My paper is filling up, and I must close by begging your pardon for speaking of the Methodists as having lost their simplicity of attire. It was true, nevertheless, for I have not seen one of the old "Simon Pure" Methodist bonnets since I have been here. But they may be as consistent as other denominations. How few of us follow in the steps of the primitive Christians.

Yours as ever,

SUSAN.

SECT. 5. — Having thus seen the factory girls at their work, so far as the picture drawn of them has been made plain to your minds, let us next look at them *in their recreations, and in their leisure employments or pursuits.*

These different particulars I put together, because it is by some results of the employments and pursuits of certain individuals amongst them, that I mean, by an example or two, to picture to you their recreations.

Briefly, let me mention, that though the leisure *employments* of the majority are no doubt like those of young women in general, employments belonging to the adorning of the person, or providing for its being done, with light, uninstructive reading, and other idle occupations, — there are considerable numbers of the factory-girls of Lowell, who are united in associations for mutual improvement; who cultivate music and other accomplishments; who attend meetings and lectures both for general information and religious edification; and there are some, and these not a few, who from time to time, write compositions for their local periodical — the Lowell Offering, — both for their own improvement, and, not unfrequently, as I believe I have proved already, for the information or gratification of others.

The production of such a work by an association of female operatives, is an incident so curious and novel, that, though Mr. Knight has placed within your reach so much information on the subject in his little book so often quoted from, — "Mind among the Spindles," — I should consider this Lecture incomplete without some account of the origin and merits of what has fairly been designated, "a phenomenon in literature."

The "Lowell Offering," as the periodical is diffidently entitled — arose out of an association of the female factory operatives called "The Mutual Improvement Society."

The first number appeared in October, 1840; and the *fourth volume* was, I presume, completed with the close of the last year, — and, summarily I may say, fair and comely, and well-filled volumes (with a broad close page of print) the four of the series are.

All the writers on America whose accounts of Lowell I have seen, unite in one general expression of surprise and approbation of the Lowell Offering. As the work of young women engaged for thirteen or fourteen hours a day (including the brief time allowed for meals) in their laboring occupation, this publication must every where excite a feeling of surprise; and yet, without any allowance for this circumstance, or for the class of society who constitute the entire body of the contributors, the Lowell Offering still stands commended, in every part of it that I have seen, for its propriety of sentiment and language, and high moral tone, for the vividness of natural feeling and good sense portrayed in numbers of its articles, and for the descriptive cleverness, and innocent playfulness, with which some of its contributions are characterized. Criticised, with a fair consideration for the circumstances under which it is produced, the highest commendations of it that I have yet met with, would, I think, be fairly justified; and criticised, on the ordinary principles of public judgment, simply on the score of its merits as a periodical, even then, whatever might be found to be below the ordinary standard of works of the periodical class, a fair quantity of matter within its pages would meet with, I doubt not, a satisfactory measure of commendation.

Judged in this latter mode of criticism, it would, I apprehend, be concluded, that the best articles are those, generally, which come before the reader with the least pretensions. Their simple, and yet graphic descriptions of the common scenes and incidents which their own eyes have witnessed; their delineations of character through the medium of sociable colloquial communication; their incidental touches of natural sensibility and feeling in speaking of their parental home and the society there; their familiar, unpretending accounts of factory ways and factory works and factory girls' pursuits, foibles, and feelings; their free communications, in letters, descriptive of their condition and impressions, to friends at home; their simple descriptions of domestic scenes and habits, and of incidents or circumstances in local history or biography; their writings of these kinds, chastened and consecrated by self-humbling reflections, or chaste and pious sentiments, as they often are, constitute, I think, the real merit, the valuable staple, of this rare species of Lowell production.

Where, on the other hand, great and profound subjects are attempted, they generally, as reasonably must be expected, fail. The fault is that of youth generally in their early efforts at essay writing. Writing freely and confidingly, to a tender mother or a loving sister, their letters are frequently good, because natural, sometimes touching, sometimes essentially interesting. But trying to write in fine language on sentimental topics, or to clothe in pompous adornment, moderate ideas of sublime subjects, they, of course, fail. That faults of this kind should appear in a variety of cases in the Lowell Offering, is only that which we should naturally look for as results of the ordinary tendencies in youthful authors. But, where there is, generally, so much to commend, and such decided marks of good sense, good taste, right feeling, and cleverness of description, I am well persuaded that an occasional hint from a judicious literary friend (which I doubt not would be well taken) for the discouragement of great attempts, and for the cherishing and encouraging of those narrative and descriptive subjects in which they really excel, would act with a beneficial tendency, not merely for keeping up the popular reputation of, but for enhancing in its literary value, this interesting periodical.

But I must not forget myself, as if the business of this evening were to criticise the "Lowell Offering," instead of telling you of the female factory operatives themselves.

You will remember, however, that I came to speak particularly of this publication, as exhibiting some of the well-looking, yes, well-tasting, fruits of their leisure pursuits. Not, be it observed, that these factory girls, are *all* learned ladies, for the contributors to a publication of this kind, are, no doubt, only a small proportion of their seven thousand. Yet the writers in the Offering are by no means few. Of this we have a sufficient evidence in the papers in Mr. Knight's selection, derived from the first two volumes only, I believe, of the Lowell Offering. For in the introduction thereto, we are told that of the thirty-seven articles given in the little volume, twenty-nine are the productions of separate contributors; and the whole of the literature of the publication, you will remember, is the production of their own class.

By the help of these writings, of the Lowell factory girls, I shall now proceed to picture to you, as I recently proposed to do, something of their *recreations.*

And for this purpose, as I can but take a single example here, I select a paper, or contribution, headed, "Leisure hours of the mill-girls." *

It is too long, however, to be read throughout. But, as the abridging of it will mar the completeness of the description and story, I will endeavor to make my extracts intelligible, by supplying, as far as my time will allow, the substance of what is omitted.

The article comprises the morning's incidents, connected with the pursuits and mutual communications of the inmates of a particular boarding-house, on a leisure day, when the mill, to which these girls belonged, was stopped for some repairs.

* Art. xix. in "Mind among the Spindles."

The principal characters introduced into the narrative (as to the domestic party) are Ann, Bertha, Charlotte, Ellinora, Fanny, Isabel, Alice, &c.

Ann, an active, well-meaning, but somewhat rash girl, "having sat at her window in the early morning 'till she was tired of watching the passers-by, started up in search of one idle as herself, for a companion in a saunter." She called first at the chamber opposite her own. The room was sadly disordered. The bed not made, though past nine o'clock. Gay dresses and mill clothes were scattered incongruously about on every article of furniture; and the inmates, in their personal appearance, were in but little better a condition. Even the pictures hanging about the walls, being paltry subjects, failed to accomplish that for which they were intended, the adorning of the room. Bertha blushed, evidently not well pleased to be caught in such disorder: the reason was, she had been hurrying all the morning to get through a novel, before her companion, Charlotte, returned.

Charlotte is described, as being little more elevated in her pursuits for enjoyment. She, though of poor parents, needing her help, was in the habit of spending her money in trifles and follies, of which an example was observed, just as Ann was leaving the room to seek elsewhere for a companion in a walk, when Charlotte came in loaded with "lemons, nuts, and cake."

Ann then tried the next chamber. Here every thing, in striking contrast with the place she had left, was in the most perfect order: the bed was shaped, the lines hanging on the walls, filled exactly with female draperies in the nicest condition. There were paintings also, so far consistent, that they were formal as the starched capes and aprons on the lines. "There was an apology for a lady. Her attitude and form were precisely of that uncouth kind which is produced by youthful artificers who form head, body and feet from one piece of shingle [a narrow thin strip of board used instead of slate for the covering of houses] and wedge in two sticks at right angles with the body, for arms. Her sleeves increased in dimensions from the shoulders, and the skirt from the belt, but without the semblance of a fold."

Miss Clark was busily engaged in making netting; and Miss Emily in making a dress. Ann makes known her wants to them, though with little hope of success. One had given herself a "stint," or task, to make five yards of netting that very day, so she could not go; the other was over head and ears in work, or she would have been "glad to."

On wishing these ladies good morning, Ann heard a loud and merry laugh from the next room, as she reached the door. But I will not spoil the cleverness and playfulness of the description that follows, by abridging it: you shall have it accurately from the book:—

It was Ellinora Frothingham's laugh which she heard; no one could mistake who had heard it once. It seemed the outpouring of glee that could no longer be suppressed. Ellinora sat on the floor, just as she had thrown herself on her return from a walk. Her pretty little bonnet was lying on the floor on one side, and on the other a travelling bag, whose contents she had just poured into her lap. There were apples, pears, melons, a mock-orange, a pumpkin, squash, and a crooked cucumber. Ellinora sprang to her feet when Ann entered, and threw the contents of her lap on the floor with such violence as to set them to rolling all about. Then she laughed and clapped her hands, to see the squash chase the mock-orange under the bed, a great russet running so furiously after a little fellow of the Baldwin family, and finally pinning him in a corner. A pear started in the chase; but after taking a few turns, he sat himself down to shake his fat sides and enjoy the scene. Ellinora stepped back a few paces to elude the pursuit of the pumpkin, and then, with well feigned terror, jumped into a chair. But the drollest personage of the group was the ugly cucumber. There he sat, Forminius-like, watching the mad freaks of his companions.

"Ha, see that cucumber!" exclaimed Ellinora, laughing heartily. "If he had hands, how he would raise them so! If he had eyes and mouth, how he would open them so!" suiting actions to her words. "Look, Ann! look, Fanny! See if it does not look like the Clark girls, when one leaves anything in the shape of dirt on their table or stand!"

Peace was at length restored among the *inanimates*.

"I came to invite you to walk; but I find I am too late," said Ann.

"Yes. Oh, how I wish you had been with us! You would have been so happy!" said Ellinora. "We started out very early, before sunrise, intending to take a brisk walk of a mile or two, and return in season for breakfast. We went over to Dracut, and met such adventures there and by the way as will supply me with food for laughter years after I get married and trouble comes. We came along where some oxen were standing, yoked, eating their breakfast while their owner was eating his. They were attached to a cart filled with pumpkins. I took some of the smallest, greenest ones, and stuck them fast on the tips of the oxen's horns. I was so interested in observing how the ceremony affected the Messrs. Oxen, that I did not laugh a bit until I had crowned all four of them. I looked up to Fanny, as I finished the work, and there she sat on a great rock, where she had thrown herself when she could no longer stand. Poor girl! tears were streaming down her cheeks. With one hand she was holding her lame side, and with the other filling her mouth with her pocket handkerchief, that the laugh need not run out, I suppose. Well, as soon as I looked at her and at the oxen, I burst into a laugh that might have been heard miles, I fancy. Oh! I shall never forget how reprovingly those oxen looked at me. The poor creatures could not eat with such an unusual weight on their horns, so they pitched their heads higher than usual, and now gave them a graceful cant, then stood entirely motionless, as if attempting to conjecture what it all meant.

"Well, that loud and long laugh of mine brought a whole volley of folks to the door — farmer, and farmer's wife, farmer's sons, and farmer's daughters. 'Whoa hish!' exclaimed the farmer, before he reached the door; and 'Whoa hish!' echoed all the farmer's sons. They all stopped as soon as they saw me. I would remind you that I still stood before the oxen, laughing at them. I never saw such comical expressions as those people wore. Did you, Fanny? Even those pictures of mine are not so funny. I thought we should raise the city police; for they had tremendous voices, and I never saw anybody laugh so.

"As soon as I could speak, and they could listen to me, I walked up to the farmer. 'I beg your pardon, Sir,' said I, 'but I did want to laugh so! Came all the way from Lowell for something new to laugh at.' He was a good, sensible man; and this proves it. He said it was a good thing to have a hearty laugh occasionally, good for the health and spirits. Work would go off easier all day for it, especially with the boys. As he said, 'boys,' I could not avoid smiling as I looked at a fine young sprig of a farmer, his oldest son, as he afterwards told us, full twenty-one."

"And now, Miss Ellinora," said Fanny, "I shall avenge myself on you, for certain saucy freaks, perpetrated against my most august commands, by telling Ann, that as you looked at this 'young sprig of a farmer,' he looked at you, and you both blushed. What made you, Nora? I never saw you blush before."

"What made you, Nora?" echoed Ellinora, laughing and blushing slightly. "Well, the farmer's wife invited us to rest and breakfast with them. We began to make excuses; but the farmer added his good-natured commands, so we went in; and after a few arrangements, such as placing more plates, &c., a huge pumpkin pie, and some hot potatoes peeled in the cooking, we sat down to a full round table. There were the mealy potatoes, cold boiled dish, warm biscuits and doughnuts, pie, coffee, pickles, sauce, cheese, and just such butter and brown bread as mother makes, bread hot, just taken from the oven. They all appeared so pleasant and kind, that I felt as if in my own home, and with my own family around me. Wild as I was, as soon as I began to tell them how it seemed to me, I burst into tears in spite of myself, and was obliged to leave the table. But they all pitied me so much, that I brushed off my tears, went back to my breakfast, and have laughed ever since."

"You have forgotten two very important items," said Fanny, looking archly into Ellinora's face. "This 'fine young sprig of a farmer' happened to recollect that he had business in town to-day; so he took their carriage and brought us home, after Nora and a roguish sister of his had filled her bag as you see. And more and better still, they invited us to spend a day with them soon; and promised to send this 'fine young sprig,' &c., for us on the occasion."

Ellinora was too busily engaged in collecting her fruit to reply. She ran from the room; and in a few moments returned with several young girls, to whom she gave generous supplies of apples, pears, and melons.

She was about seating herself with a full plate, when a new idea seemed to flash upon her. She laughed and started for the door.

* * * * *

Fanny turned to one of the ladies who had just entered.

"Where is Alice?" said she. "Did not Ellinora extend an invitation to her?"

"Yes; but she is half dead with the *blues*, to-day. The Brown girls came back last night. They called on Alice this morning, and left letters and presents from home for her. She had a letter from her little brother, ten years old. He must be a fine fellow, judging from that letter, it was so sensible and so witty too! One moment I laughed at some of his lively expressions, and the next cried at his expressions of love for Alice, and regret for her loss. He told her how he cried himself to sleep the night after she left home; and his flowers seemed to have faded, and the stars to have lost their brightness, when he no longer had her by his side to talk to him about them. I find by his letter that Alice is working to keep him at school. That part of it which contained his thanks for her goodness was blistered with the little fellow's tears. Alice cried like a child when she read it, and I did not wonder at it. But she ought to be happy now. Her mother sent her a fine pair of worsted hose of her own spinning and knitting, and a nice cake of her own making. She wrote, that, trifling as these presents were, she knew they would be acceptable to her daughter, because made by her. When Alice read this she cried again. Her sister sent her a pretty little fancy basket, and her brother a bunch of flowers from her mother's garden. They were enclosed in a tight tin box, and were as fresh as when first gathered. Alice sent out for a new vase. She has filled it with her flowers, and will keep them watered with her tears, judging from present appearances. Alice is a good-hearted girl, and I love her. But she is always talking or thinking of something to make her unhappy. A letter from a friend, containing nothing but good news and assurances of friendship, that ought to make her happy, generally throws her into a crying fit, which ends in a moping fit of melancholy. This destroys her own happiness, and that of all around her."

"You ought to talk to her, she is spoiling herself," said Mary Mason, whose mouth was literally crammed with the last apple of a second plateful.

"I have often urged her to be more cheerful. But she answers me with a helpless, hopeless, 'I can't, Jane! you know I can't; I shall never be happy while I live.'"

* * * * * * * * *

"Would it improve her spirits to walk with me?" asked Ann.

"Perhaps it would, if you can persuade her to go. Do try, dear Ann," answered Jane. "I called at Isabel Greenwood's room as I came along, and asked her to go in and see if she could rouse her up."

Ann heard Isabel's voice in gentle but earnest expostulation, as she reached Alice's room. Isabel paused when Ann entered, kissed her cheek, and resigned her rocking-chair to her. Alice was sobbing too violently to speak. She took her face from her handkerchief, bowed to Ann, and again buried it. Ann invited them to walk with her. Isabel cheerfully acceded to her proposal, and urged Alice to accompany them.

"Don't urge me, Isabel," said Alice; "I am only fit for the solitude of my chamber. I could not add at all to your pleasure. My thoughts would be at my home, and I could not enjoy a walk in the least degree. But, Isabel, I do not want you to leave me so. I know that you think me very foolish to indulge in these useless regrets, as you call them. You will understand me better if you just consider the situation of my mother's family. My mother a widow, my oldest brother at the West, my oldest sister settled in New York, my youngest brother and sister only with mother, and I a Lowell factory girl! And such I must be,—for if I leave the mill, my brother cannot attend school all of the time; and his heart would almost break to take him from school. And how can I be happy in such a situation; I do not ask for riches; but I would be able to gather my friends all around me. Then I could be happy. Perhaps I am as happy now as you would be in my situation, Isabel."

Isabel's eyes filled, but she answered in her own sweet, calm manner;

"We will compare lots, my dear Alice. I have neither father, mother, sister, nor home in the world. Three years ago I had all of these, and every other blessing that one could ask. The death of my friends, the distressing circumstances attending them, the subsequent loss of our large property, and the critical state of my brother's health at present, are not slight afflictions, nor are they lightly felt."

Isabel's emotions, as she paused to subdue them by a powerful mental effort, proved her assertion. Alice began to dry her tears, and to look as if ashamed of her weakness.

"I, too, am a Lowell factory-girl," pursued Isabel. "I, too, am laboring for the completion of a brother's education. If that brother were well how gladly would I toil! But that disease is upon his vitals, which laid father, mother, and sister in their graves in one short year. I can see it in the unnatural and increasing brightness of his eye, and hear it in his hollow cough. He has entered upon his third collegiate year, and is too anxious to graduate next commencement, to heed my entreaties or the warning of his physician."

She again paused. Her whole frame shook with emotion; but not a tear mingled with Ann's, as they fell upon her hand.

"You see, Alice," she at length added, "what reasons I have for regret when I think of the past, and what for fear when I turn to the future. Still I am happy, almost continually. My lost friends are so many magnets, drawing heavenward those affections that would otherwise rivet themselves too strongly to earthly loves. And those dear ones who are yet spared to me, scatter so many flowers in my pathway, I seldom feel the thorns. I am cheered in my darkest hours by their kindness and affection, animated at all times by a wish to do all in my power to make them happy. If my brother is spared to me I ask for nothing more. And if he is first called, I trust I shall feel that it is the will of One who is too wise to err, and too good to be unkind.

"You are the most like my mother, Isabel, of any one I ever saw," said Ann. "She is never free from pain, yet she never complains. And if 'pa, or any of us, just have a cold or headache, she does not rest till 'she makes us well.' You have more trouble than any other girl in the house; but instead of claiming the sympathies of every one on that account, you are always cheering others in their little half-imaginary trials. Alice, I think you and I ought to be ashamed to shed a tear, until we have some greater cause than mere home-sickness, or low spirits."

"Why, Ann, I can no more avoid low spirits than I can make a world!" exclaimed Alice in a really aggrieved tone. "And I don't want you all to think that I have no trouble. I want sympathy, and I can't live without it. Oh that I was at home at this moment!"

"Why, Alice, there is hardly a girl in this house who has not as much trouble, in some shape, as you have. You never think of pitying them: and pray what gives you such a strong claim on their sympathies? Do you walk with us, or do you not?"

Alice shook her head in reply. Isabel whispered a few words in her ear—they might be of reproof, they might be of consolation—then retired with Ann to equip for their walk.

The remainder of this story, which is full of interest, I shall only have time to sketch in outline. The young friends, Ann and Isabel, now at length commence their walk,—Ann, exulting in the enjoyment, happy as a liberated bird that has been captive, and then, entering into conversation with Isabel about the unhappy associate they had just left. Ann expresses her wonder that Alice should not, in the realization of her various consolations, and by participation in the exhilarating influence of recreation in the open air, be happy, but is reminded by her friend 'that the minds of different persons are like instruments of different tones; the same touch thrilling gaily on one, mournfully on another.' Taking up the idea, Ann expresses her assent, observing, 'that she knows that different minds may be compared to the same instrument *in* and *out* of tune.' 'Alice, she had heard say, loved to indulge this melancholy; that she loved to read Byron, Mrs. Hemans, and Miss Landon, until her heart was as gloomy as the grave.' "Isn't this strange," she asks, "even silly?" "It is most *unfortunate*, Ann," was her considerate friend's reply. "Isabel," renews the enquirer, "you are the strangest girl! I have heard a great many say that one cannot make you say anything against anybody; and I believe they are correct. And when you reprove one, you do it in such a mild pretty way, that one only loves you the better for it. Now, I smash on, pell-mell, as if unconscious of a fault in myself. Hence I oftener offend than amend."

She solicits Isabel to be to her a sister and counsellor,—and elicits the reflections of a well-ordered and devotional mind, expressive of her own inability to do anything but by the Divine assistance and the guidance of His Spirit, and recommends *prayer*—confiding, devout prayer,— as the special means appointed for the realizing of heavenly guidance and blessing. The conversation continues in the like serious and useful strain, until, on returning from their walk, they recognize poor Alice still sitting in unaltered depression. The true remedy, in religion, is then spoken of, and Isabel musingly and judiciously adds, "if she can be induced to read Byron less, and her Bible more; to think less of her own gratification, and more of that of others, — Alice may yet be happy."

Mutual explanations and apologies follow, on the return of the friends to their associates. The sound principles developed in the previous conversation, as to the religious and moral correctives of defects of character, are stated as being acted on subsequently, and with, for the most part, encouraging success. Alice's melancholy subsides, and a better condition of mind obtains; Ann's faults, in considerable degree, are counteracted. As for the others, I give the result in the words of the story: —

The Clark girls do not, as yet, coalesce in their system of improvement. They still prefer making netting and dresses, to the lecture-room, the improvement circle, and even to the reading of the "Book of books." So difficult is it to turn from the worship of Plutus!

The delusion of Bertha and Charlotte is partially broken. Bertha is beginning to understand that much reading does not naturally result in intellectual or moral improvement, unless it be well regulated. Charlotte is learning that "to enjoy is to obey;" and that to pamper her own animal appetites, while her father and mother are suffering for want of the necessaries of life, is not in obedience to Divine command.

And, dear sisters, how is it with each one of *us*? How do *we* spend our leisure hours? Now, "in the stilly hours of night," let us pause, and give our consciences time to render faithful answers.

SECT. 6. — To the last particular proposed for consideration, about the Factory-girls of Lowell, I now come to speak, which is one of deep and interesting importance: viz. — *On their general character and moral condition.*

This, it is evident, must be a somewhat difficult, or at least delicate subject to discuss — particularly as all estimates of moral character of the nature of this we purpose, must be made with reference to some well known class by way of comparison, — and comparisons, if either class shall fall greatly below the other, may become unpleasing, if not offensive.

My hope, however, is, that no just ground of offence or objection wiil be found on either side. Having no possible motive for depreciating my own countrywomen, neither any reasons for flattering our transatlantic sisters, — my endeavor will be to picture to you, fairly and impartially, according to the result of very much inquiry and consideration, what I believe to be *their* character and moral condition, as a class ; and if you, yourselves, should come to the conclusion, that they are in advance of us in what renders the female character lovely, or any character really estimable, — then my earnest prayerful desire would be, that my own young countrywomen, of the operative classes, might give a thoughtful consideration to this case of Lowell, so that, by God's blessing, they might profit thereby.

Long before I thought of visiting the United States, I had heard a surprising account of the superior character of the female factory operatives of Lowell. In certain books of Travels, I had seen them described as something so extraordinary in character and mind, in purity and religion, that but little knowledge of human degeneracy was sufficient to justify the questioning of the accuracy of such representations. The descriptions, indeed, appeared evidently those either of very partial witnesses, or of those but little acquainted with the common nature of humanity ; or else of writers so singularly undiscriminating as to fancy that outward appearances and pretensions were real proofs of the condition of the heart.

My own impression, therefore, was, that these females were greatly overrated, and the descriptions of their moral character excessively exaggerated. Learning, from the unerring word of Truth, the deceitfulness, corruptions, and even desperate wickedness of the unrenewed heart of our species; (Jerem. xvii. 9,) and knowing, both from the humiliating consciousness of what passes within the heart, personally, and from the sad faults and failures of many, in outward rectitude and virtue, of whom we had thought well, — no place was left for the extravagant notion that the young persons referred to were either different from others in their infirmities, or free from the failings and vices common to humanity.

At the same time, *comparatively* considered, these young women, it was assumed, might, from causes not equally operating with ourselves, have attained to a higher moral condition, (as they clearly have to intellectual superiority,) than the like class of females either in this or other countries.

This was an inquiry, therefore, of much interest and importance; for if the assumed superiority were proved to exist, it might naturally stimulate our Christian philanthropy to seek to make the example profitable.

These preliminary remarks I have been induced to make, to show that my inquiries on this subject were likely to be made with caution; and, having heard not only doubts expressed whilst I was in America about the prevailing ideas thereon, but found that severe attacks upon these very girls had been made in certain respectable Journals of that country, my previous caution was naturally increased.

The effect of all this was, that I expected to find the female operatives here not very greatly different, in their actual moral condition, from what we find them elsewhere;

but, as you will soon learn from what I have further to say on this subject, this expectation did not agree with what I now believe to be the fact.

The position in which a majority of these young women are placed, — as strangers from the country, separated from their natural protectors, and left almost entirely to their own discretion and moral self-dependence, — is one singularly exposed to risk and temptation. Nor, under ordinary influences, would their being congregated in such considerable numbers together in their sleeping-rooms and boarding-houses, by any means tend to their greater security.

But from all the information I could obtain, tested by all available means, and received with all reasonable caution, I have been led to infer that their character and moral conduct are unusually high. Not, indeed, that the impressions calculated to be conveyed by certain published statements, already referred to, were, or could reasonably be expected to be, verified; not that instances of failure did not occur, nor that such were so rare as to be all but unknown; not that these factory-girls were free either from the passions or the vices of their class and sex elsewhere; but still, that, after every allowance, their *general* moral character, as an entire class, appeared to be left in a very superior and commendable position.

This, however, you should observe, is not altogether peculiar to the young women of Lowell; it is the character very generally maintained, speaking comparatively with reference to other countries, of the young women of New England. So that whilst we infer, that, in no other part of the world is so high a general standard, among the females of a manufacturing population, known, as it is in the place of which we speak; it is only intended to be intimated, that, with respect to the rest of the young women in the like rank of life in New England generally, the standard, with the female operatives of Lowell, may be considered to bear a fair comparison. More than this, we would not venture to assume.

And from information derived from sources unconnected with the factories—such as from those who watch for their spiritual interests and care for their souls—the same conclusion was arrived at. The ultimate reply to my inquiries on this subject was to the effect,—that "we are very thankful if we can keep up the standard of morals so as to be equal to what prevails amongst other young women of New England."

That this indicates a fair judgment on the point under consideration, is verified, I think, by the result of searching investigations which naturally rose out of the severe attacks on the character of the young women of Lowell, already referred to. Dr. Bartlett, a medical practitioner of Lowell, in no way interested, he tells us, in any of the factories, has come forward, in a manly, and, I think, satisfactory pamphlet, in "vindication of the character and condition of the females employed in the Lowell mills." The statistics, introduced for refuting the charges that had been publicly made, are very interesting, and on some of the points eminently conclusive.

The result of his views, as to their moral condition, Dr. Bartlett thus states:—"The great body of the girls are as virtuous as the female population of any part of New England; they have an honest and conservative pride in the preservation of their character and respectability as a class,"—a fact, indeed, which the Lowell Offering repeatedly indicates. And so far from their being injured by their removal from the country to work in the factories, this author is of opinion, after a close observation of them for many years, that it is the very reverse; 'the aggregate change wrought in their moral character and condition being, as he believes, eminently happy and beneficial.' "The great preponderance of influence," he adds, "is enlightening, elevating and improving—not darkening, debasing, and deteriorating. Their manners are cultivated, their minds are enlarged, and their moral and religious principles are developed and fortified. Hundreds and hundreds of these girls will long live to refer the commencement of their best and highest happiness to their residence in this city." [p. 21.]

These conclusions, however, may or may not be *perfectly* accurate. And we might still hesitate as to where, in the scale of moral excellency, or otherwise, the female operatives of Lowell should be placed.

But there are certain facts, to which we can refer, that may considerably aid our candid inquiries in this matter. Some ten or twelve years ago, when the factory-girls at Lowell were about 4000 in number—little more than one-half of what they now amount to—a visitor at that city informs us, that there then were about 1000 who were "united to Christian Churches" (as members) and about the same number who were in regular attendance on the means of grace.*

In 1839, Dr. Bartlett informs us, that there was an aggregate (by estimation) of 6000 souls, nearly one-third of

* Narrative of a visit to the American Churches, by Drs. Reed and Matheson, vol. I. p. 417.

the entire population of Lowell, connected with their Sunday Schools; that under one union of ten schools, there were 5369 scholars and teachers, and that, of the female part, in these schools, comprising about three-fourths of the whole, "a large proportion were above fifteen years of age, and consisted of girls employed in the mills." The Report of the union quoted by Dr. Bartlett, further states—'that more than 500 of these scholars had, during the previous year, become personally interested in practical piety, and more than 600 had joined themselves to the several churches.' [p. 18.]

From what has now been stated, a fair estimate, I hope, may be formed of the general character of the factory-girls, as a class. The statements, it will be observed, have been general, and the comparisons general. Otherwise, we could not have concluded in such decided terms. For, as to *particular* cases, there are examples in our own country, where a deep sense of moral responsibility on the part of employers, has urged to vigilant, judicious and praiseworthy efforts, for the checking of vicious habits, and the encouragement of those of morality and piety, among the females employed by them, which have resulted in the most satisfactory consequences.

As to the character of the factory-girls of Lowell, otherwise considered than by a fair and probable medium—there are, no doubt, the ordinary varieties of good and bad, moral and vicious. As to the extreme contrasts, however, of the religious and educated character—such as the avowed infidel and the stupidly ignorant—these, I apprehend, hardly exist among the females of Lowell.

As to their character in things not pertaining to morals,

we find, they have their foibles, and their errors too. The gaiety and expensiveness of the dress of many of them, has been noticed more than once.

Mr. Dickens, in his American Notes, under the authority only, indeed, of "it is said," mentions "that on the occasion of a visit from General Jackson or General Harrison to this town, he, the General, walked through three miles and a half of these young ladies, all dressed out with parasols and silk stockings."

This might or might not be true; but I quote the said-to-be-true incident, because it applies to what is no doubt true of the fondness for dress, and the too free indulgence of expense, among the masses of these young females.

In respect to the superior intelligence, indicated by the selections I have read to you from the Lowell Offering, — this is only to be understood as pertaining to a portion of the female operatives; and these, capable of such literary efforts, of course form but a small proportion of the whole body. Yet, after all — after every allowance for over-statements or partial statements, for exceptions to the general moral character, or superior intelligence, or great respectability — these factory-girls do greatly surprise and interest us.

CHAPTER IV.

INQUIRY INTO THE CAUSES OF THE SUPERIOR MORAL AND INTELLECTUAL CONDITION OF THE LOWELL FACTORY-GIRLS.

CONCLUDING as we must do, if the foregoing researches and observations be at all correspondent with the facts, that the intellectual, moral and religious condition of the female operatives of Lowell is of a very unusual and superior order, — the inquiry, as to how these beneficial effects are produced? becomes as interesting as it is important.

A variety of causes, no doubt, contribute to the general result. Of these we may notice, —

SECT. 1. — *The general superior tone of moral principle, and propriety of behavior, prevalent among the young women of the operative classes in New England.*

This point which has been just touched upon already, I merely notice again in this place; as the subject of the general moral characteristics of the country belongs to another lecture in my proposed series. It may be sufficient for the present, therefore, to assume the fact, that, comparing the females of America of the operative classes (those of the north-eastern States especially) with the corresponding classes in the several countries of Europe generally — such superiority is generally admitted, I believe, decidedly to exist.

SECT. 2.—*The universal prevalency of education among all classes, connected with the general respect for the Bible and religion.*

Education, alone, considered as the mere cultivation of the intellectual part of man, without regard to the moral faculties, will have the effect of altering the external character, and rendering less odious and disgusting, in outward appearances, the vices to which our species is prone. But secular or heathen intelligence, however highly cultivated, will have no effect on the actual condition of man's moral constitution. For secular knowledge has no relation to, nor affinity for the moral and spiritual perceptions of our nature, and therefore can neither move nor improve them.

Education, however, so connected, as in New England it prevalently is, with respect for the Bible, the Lord's day, and religion, and with the usage of the people of carrying out such respect into practical efforts, becomes, virtually, a religious education, and therefore of a moral tendency. To have its full effect, indeed, in moral and spiritual renovation, and to subserve the object Divinely designed, the *species* of religion embodied in the education *must be sound, and essentially that of the Gospel;* but though defective in this, as some theological systems vitally are, yet any form of Christianity, sincerely professed, unsaving though it be as to a future state, may exert on human actions, temporally considered, a restraining moral influence.

In the case under consideration, the education of the masses, whether positively or negatively religious, is always associated, by Sunday school teaching, or the religious conventionals of the people, with religious directives. Thus the influence yielded by education obtains a moral application. The knowledge, which abstractly is *power*, is diverted from nature's guidance, tending to evil, into heavenly guidance, influential in good.

The original education of the factory-girls is not higher, generally, than that of others in the same country; and it can by no means be equal, in the case of large numbers from the country districts, to what a like class would have received, if partaking of the advantages afforded by the effective common schools of their principal cities.

Respecting the *generality* of an elementary education, a single fact—stated to me by the manager of one of the companies' factories—will suffice to be mentioned. 'All the girls in the mill under his care' (this gentleman assured me) 'were generally very well educated. An examination which had recently taken place,' he said 'in respect to 800 girls, belonging to the four mills of his corporation, showed, that there were only 43 out of that number who did not write their names legibly and tolerably well; of these, 40 were Irish, and 2 others *not* natives!'

SECT. 3.—*The considerable wages which they receive*, is not without its influence.—This circumstance, of itself, cannot, of course, exercise any moral influence. But, negatively, a fair and adequate means of subsistence and comfortable maintenance, is essential for preservation from the peculiar snares of poverty, and from that reckless abandonment of self-respect, and even of female virtue, which too naturally follow in the train of wretchedness and worldly degradation. A wretched population, as to poverty and want, can hardly be a virtuous population.

Religiously considered, a poor man, as well as one of any other class, may be a God-fearing man — even the *beggar* Lazarus, may be a son of Abraham and a privileged saint. But *economically* and *morally* considered, *extreme poverty* presents great and peculiar snares. This the pious Agur intimated in his brief, but instructive prayer, — "Give me neither poverty nor riches ; feed me with food convenient for me ; lest I be full, and deny thee, and say, who is the Lord ? or lest I be poor, and steal, and take the name of my God in vain." (Prov. xxx. 8, 9.)

The ample remuneration, therefore, received by the factory-girls, at Lowell — though with many it ministers, no doubt, to that love for dress and expensiveness which they too often carry to a very unfitting excess — both secures them from the peculiar temptations incident to poverty, and enables this cherished feeling of independency to take the line of self-respect, and thus to aid, under the guidance of their general religious knowledge, in producing a higher practical standard of moral virtue.

SECT. 4. — Another favorable influence is *the watchful consideration and moral care, for the young women, taken by their employers and others.*

Dr. Bartlett, speaking on this point, specifies among the favorable influences—"the example and watchful care and oversight of the boarding-house keepers, the superintendents, and the overseers. The moral police," he says, "of all the establishments is vigilant, active, and rigid. While industry and good conduct are respected and rewarded, no violations of the excellent and judicious rules of the corporations, and no improper or suspicious conduct meet with any indulgence or toleration. It is only by maintaining an unsullied and unimpeachable character, that a girl can retain her situation in the mill, and when dismissed for any impropriety from one establishment, there is no possibility of her getting a place in any others."

The overseers, it is stated, are generally married men — carefully selected, it would appear, not merely with regard to their qualifications as men of activity and intelligence, but as to their fitness as moral agents.

In a chance case, Dr. Bartlett applied for some information on the subject of overseers to one of the superintendents of a corporation, who informed him — 'that thirty overseers, and as many second overseers, were employed in the mills, and the various departments connected with them, under his charge. Of these overseers, all but one were married men, with families, the one forming the exception was about to be married. Of the second overseers, being younger men, upwards of two-thirds were married, and several others about to be married. Twenty-six of the overseers of the two classes were members of some regular church, (that is, communicants,) and four of them deacons.' And, as a body, the overseers referred to are further described by the superintendent, as challenging comparison with any class of the community in all the sterling requisites of character, — "in intelligence and practical good sense, in sound morality, and as active, useful, exemplary citizens."

It should be added, which I can do safely from my own knowledge, that some, probably very many, of the principal proprietors of the factories of Lowell, are men highly distinguished among their countrymen for piety and benevolence, — and who, therefore, exercise a watchful and beneficial influence over the moral tone of the operatives they employ.

I would not, however, have you suppose, that all is or can be unexceptionably good. All that I would confidently state, in respect to the moral influence derived from these sources, is, that, as a whole, it is decidedly favorable and influential. Nor would I wish to convey the impression, that all the influence that is exerted springs from pure benevolence or sound religion. No doubt, a variety of motives, with a fair share of self-interest, concur in producing the general result. And one influencing consideration, which may be unquestionably adduced, is the fact, that the keeping up of the moral tone, and high character of these establishments, is of great importance in securing their prosperity. Were the moral character of the female operatives otherwise, the respectable classes in the country, who contribute so mainly to the supply of hands, would no longer trust their daughters to a demoralizing influence; the sources, therefore, being walled in, the stream would cease to flow from that direction. As it is, the supply is often inadequate to their requirements, and at this very time, I believe, the growth of these manufactories is checked by the difficulty of obtaining the requisite amount of hands. Did they, however, admit among them an inferior class, as to moral tone and conduct, the present operatives, as a body, would, I have no doubt (as the next consideration will show you), immediately leave their employment and return to their country homes. So that various considerations, besides the impressive sense of moral obligation and responsibility, of right and religious feeling on the part of the managers and corporations, conduce to the promotion of the same praiseworthy object. Yet, whatever measure of self-advantage may be engaged in this object, the general benefit is not hindered.

As a result, indeed, of well-doing, self-advantage comes in as a scriptural encouragement.*

But this watchful care is not confined to the departments of society interested, pecuniarily, in the manufactories of Lowell. These are aided and supported in a most efficient manner by the ministers of religion, in their place and province, and by the zeal of the managers and teachers of their numerous Sunday schools, and other institutions connected either with religious instruction, or moral improvement. Altogether, a variety of instrumentality in aid of the stranger-girls, appears to be at work,—all acting harmoniously and effectively for the conservation of moral principle, and the promotion of piety among them.

SECT. 5.—Lastly, I have to notice, as an influence peculiarly beneficial to the Lowell factory-girls,—*The self-consideration, and regard for moral character, of the female operatives themselves.*

Everywhere where Lowell is spoken of, or its factory system, in practice, described, this commendable feeling of the female operatives is distinctly indicated. Dr. Bartlett considers "the jealous and sleepless watchfulness, over each other, of the girls themselves," as the pre-eminent influence in the production of their high tone of moral character. This influence may be overrated,—as I believe their careful training, generally, in reverence for religion, the Bible, the Sabbath, and for the external character befitting a Christian profession, to be the pre-eminent sources of their moral superiority,—yet it no doubt operates as a most efficient handmaid to the practical carrying

* Prov. xix. 17: Ps. xli. 1-3: cxii. 2, 3: 1 Tim. iv. 8.

out of sacred precepts otherwise inculcated. But the results, as they are stated by Dr. Bartlett, I presume to be essentially correct. His statement (partially quoted already) is, that the factory-girls "have an honest and conservative pride in the preservation of their character and respectability as a class;" that "they will not suffer either overseer or superintendent to be indifferent to this character with impunity:" and that "their censorship is despotic, so that no superintendent nor overseer could keep his hands, if he failed to remove an obnoxious or suspected individual, when the finger of censorship has pointed her out."

My personal inquiries, both at Boston and Lowell, gave, very nearly, the same results. A highly respectable superintendent represented 'the moral character of the girls, and their good order, as almost unexampled; and the general tone of character and feeling, as being so high in the works, that any girl becoming disreputable would not be associated with, but obliged to leave; and that an instance of an illegitimate child was hardly known.'

I have seen it elsewhere stated, but did not happen to ascertain the existence of the practice at present, that "the steady girls who work in a mill, band together as a sort of a club, and keep up a sense of honor through the establishment. If any one is suspected of bad conduct, she is reprimanded and suspended," (a course, however, for mere suspicion, if such be really practised, rather too hard,) "and if bad conduct is proved against her, she is reported to the managing party, and a petition is presented for her removal. In an unquestionable case they would leave the mill if the prayer of the petition was refused."*

* Reed and Matheson's visit, &c., vol. I. p. 417.

CHAPTER V.

SUGGESTIONS FOR THE IMPROVEMENT OF THE CONDITION OF *OUR* FACTORY OPERATIVES, — ESPECIALLY THAT OF FEMALES.

In the foregoing chapters I have endeavored to describe, with all the impartiality and completeness which the means within my reach enabled me, the state and condition, economically, intellectually, and morally, of the factory girls of Lowell. By no means, however, should I feel satisfied with myself to leave a subject of so much interest and importance to the chance of some practical use being made of it by others, without the endeavor, on my own part, to do something towards the turning of an example, in which there is so much that is worthy of imitation, to account. And whilst the public position which I still retain among you (for the present time at least), demands of *me*, I conceive, such practical application of the subject-matter of my Lectures, with a view to our general benefit as a manufacturing population,—*you*, I trust, will see it to be a duty, on your part, to receive, with a candid and self-appropriating mind, the practical suggestions which I have now to offer for your consideration.

Whilst the case of Lowell, however, will afford us some leading and important suggestions towards the object we have in view,—that case, it will be easily perceived, has its peculiarities; such peculiarities, therefore, must prevent what is done there becoming in all respects a model for us.

One great and striking difference, in the present positions of England and America, is found in the history and rise of our respective manufacturing establishments. In England, for the most part, they consist of enlargements of old, and successions of new manufactories, within a previously existing manufacturing town. But in America, the manufactories consist altogether of new, or at least of very modern erections; and these are not unfrequently planted, as in the case of Lowell, not as grafts upon the stock of a town venerable in age, but as the original occupants of the ground, and as the stock on which the town itself is grafted.

Hence the American manufacturing towns possess, in many cases, the advantage of a peculiar unity of construction, each part, like the mass of machinery in a factory, having an essential relation to, or connection with, the whole. There is no previously existing idle and profligate population to inoculate with its viciousness the incoming country females; nor is there already there any unhappily pauperized population to depress with its burdensomeness the enterprise of the manufacturers. They have not, as with us it commonly happens, to contend with pre-existing difficulties, with inconvenient or disadvantageous pre-occupations; but have, or in most cases may have, a clear and unembarrassed field for their manufacturing enterprises; a field wherein they can build on a plan to their liking, adopt the best models, introduce the most perfect and efficient machinery, and then order the principal contingencies, requisite for carrying on their works, on some commodious general plan.

Again, find we a want of analogy in the two countries in several other circumstances, more or less peculiar to the manufactories of the United States, which may just be noticed; such as, in the high prices they obtain, through their protective tariff and other causes, for their manufactured goods consumed at home; in the ample wages which, from the general scarcity of laborers, the operatives may demand, and which the remuneration in high prices enables the manufacturer to pay; in the cheapness of their board and lodging, yielding more clear gain and saving to the prudent operatives; and in the superior and more independent class of persons, who, by reason of these latter advantages, are induced to engage in the business of the factories.

These differences, besides others, betwixt the circumstances of our manufacturing population and that of the United States, it is but fair to keep in view, when especially, the comparing of the factory population in the two countries may leave us in so unfavorable, if not humiliating, a position. Nevertheless, admitting all these differences, with the full weight of advantage they give to our transatlantic factory workers,—there yet remain several very important circumstances and practices in the factory-system of Lowell, of a highly moral and beneficial tendency, which, with some attention and a little self-denial, would be found abundantly fitted both for our emulation and imitation.

To the suggestions, arising out of these commendable and appropriate practices, concluded by others which have occurred to me as adapted for our peculiarities of condition,—I now proceed to direct your attention; and for the sake of distinctness, I shall place the leading suggestions in separate sections.

SECT. 1. — And, first, let me suggest to you, my young friends, that both your interest and duty unite in soliciting you *to emulate your American sisters in their zeal in the acquiring of education.*

A redeeming element, of great importance and benefit in our present factory system, is the enforcing of a certain measure of school-attendance on the younger children. This is, at once, the praiseworthy result of the paternal care of the government, and the practical expression of its conscious responsibility.

Towards the rendering of this benevolent regulation of the legislature the more effective here, I have done my endeavor, to the utmost of my ability, for the erection of schools (already amounting to several in number) and the employment of superior teachers, within various districts of this town and parish.*

But the early age at which our young people are permitted to go on full time, greatly limits the benefit of the best previous education, and so operates too, that, if no pains be taken to keep up what has been acquired, much of the instruction will necessarily be lost.

Hence the importance to yourselves to emulate the praiseworthy example of the factory girls of America, in your zealous endeavors, not only to keep up such education as you may have already received, but also to improve your minds by the acquisition of additional knowledge.

* The number of National Day Schools referred to, built or opened within the last four years, reckoning each separate room in which the children are confided to a distinct master or mistress, (and including previously existing schools fitted up and brought into operation,) amounts to ten, and the amount of accommodation therein provided (reckoning six square feet per child) is sufficient for 2470 children.

You have been particularly informed about the efforts they make for self-improvement, and for raising themselves in consideration as rational beings, and thus advancing their place and condition in life. It may not be practicable for you all to do what they do, nor to attain to what they have attained; but, most of you, if your hearts were set upon it, might do a great deal in the way of your own improvement. It is quite in your power, with the respectable earnings which many of you have, to obtain the few unexpensive books that are necessary for enabling you to keep up your "schooling," and to improve yourselves at home. It is quite in your power, too, to employ your minds, for at least a part of the time you are at work, in thinking over the subjects you had previously read for improvement, rather than leave the current of your thoughts to run unchecked in idle, unprofitable, or even worse, reveries.

But a better help towards your mental improvement would be, the attending for an hour, if but three or four times a week, an evening school or other means of instruction, so as perseveringly to add to your knowledge, and to seek, by this and every godly proceeding, to raise your condition.

True it is that you are long employed in your factory-labors, for longer time, I believe, than is fitting or beneficial either for yourselves or others; but still you have not longer hours than the factory operatives of Lowell. So that that which they do for self-improvement cannot be altogether impossible with you.

And if a sufficient number of you were so disposed, happy should I be to assist you, in so laudable and important an object, by every means in my power. Both rooms and teachers, I should be able to make available, so as to provide for the instruction of as many as might be disposed to attend. And, if those who have a desire for improvement, would steadily cultivate that desire, it would not be difficult, by the means and accommodation, such as we have in the Church Institution, and our several school-rooms, and class-rooms, to aid you in any measures of this kind, and that, I believe, in an effective manner, for so ennobling a purpose.

Happy will it be for you and your parents, now, whilst you are occupied as factory operatives; happy will it be for you when you shall become wives and mothers; happy will it be for your husbands and children; happy will it be for the town in which you labor, and for the people among whom you associate, when your zeal shall successfully urge you to emulate, by religious attainment and intellectual cultivation, what strangers have so admired in many of the factory girls of Lowell.

SECT. 2.—In order to the attainment of such a condition, or rather to the attainment of a condition befitting your capabilities and your sex, let me also suggest to you *the careful cultivation of a high tone of chaste and moral principle.*

In the case of Lowell, you will have remarked, that whatever others have done for the factory girls there,—that whatever care and watchfulness may have been bestowed on them by superintendents and overseers, ministers and teachers,—yet all would not do without a watchful care over themselves.

And such, remember, is the general fact, that no care of others will be sufficient to keep you respectable and virtuous, without your own care. A female, who has no value for reputation or character, is already a lost woman. One who has no self-respect, will gain none from others. Where, however, there is a proper self-respect, *character* will be valued beyond almost any other worldly consideration, except the possession of that true chastity, the produce solely of religion in the heart, which the possession of character always supposes. So *precious*, remember, is the preservation of female purity, that a single error herein is a blemish to respectability for life. There is no repentance for that failure, that can restore a woman to the same place in society, as she might and ought to have possessed, though it be sought with many tears. Well would it be for our general population, if this were more felt, so as to act as a salutary check against the too prevalent and lamentable profligacy of our country.

In your care for yourselves, as you would reasonably hope that your care should be successful, let me press on you the great importance of your choice of companions, and your most careful consideration of the character of your associates. The next of kin to personal viciousness, is the associating with profligate companions! The keeping company with those of loose or doubtful characters, provides both for a participation in their disreputableness, and the inheriting of their vices. The associating, or having anything to do with such persons is essentially demoralizing, destructive of self-respect, and commonly ends in the loss of character.

The conduct of the Lowell girls on this point, as I have particularly noted, is very strict and worthy of imitation. A case named to me by one of the factory-proprietors, may be given as an example of their peculiar strictness. A change, I forget for what cause, had to be made in the keeper of one of the corporation boarding-houses. A person was selected, or named for the appointment, a widow, on whose character, however, about three years before, there had rested some cloud or suspicion. The girls applied to the overseer and told him, 'that if this individual were placed in the house they should *all leave*.' The reply was, 'that there was no just ground, as the overseer believed, for the report that had prevailed, and that she would be able, he thought, to explain the matter to the satisfaction of the boarders.' But this did not satisfy their anxious scruples. They declared 'that if she were appointed they could not remain. Reports,' they said, 'of a disreputable nature, had gone abroad concerning this individual; and *as their character was every thing to them*, they must not risk damage to it by boarding with one that was suspected to be immoral.' The result was, another appointment was made.

This strictness might here amount to a hard case, provided the person were really innocent; but I mention it to show their delicate sense of the importance of character in any person with whom they had to be associated.

Now should not this, my young friends, be an example for you? You, who have parents residing sufficiently near your work, will naturally reside with them. It may be for good or evil, according to their personal character. But your lot herein being providentially appointed, you may hope, under a careful seeking of the Divine protection, to be kept from the example, if it be evil. But you who have not parents resident with whom you may abide, are bound, as you care for your welfare in body and soul, to look very carefully and vigilantly into the character of the persons in whose houses you lodge, and into the character too of those who lodge with them. For want of this, what numbers of young persons, unsuspectingly trusting themselves in the houses of unprincipled strangers, and encouraged to do so, perhaps, by the temptation of cheap board and lodging, have fallen a prey to the designs of the wicked, and have been utterly ruined. Bradford, we know, in common no doubt with every similar town, has presented cases of this kind of the most lamentable nature. Take heed to yourselves, then, as to those you associate with, where you work, and where you lodge. Your happiness, both in this world and for ever hereafter, may be sacrificed by an error in either of these particulars.

But, need I say to you, that all your concern for yourselves, without the Divine blessing, may fail to keep you safe? In regard to character, let me remind you, that the only real security is the possession of true religion. Without religion, there very possibly may be, in your rank of life as in that of those above you, a general propriety of manner and moral conduct; but there cannot be inward purity, real excellence, or perfect integrity, without religion in the experience of it, and godliness in the power of it.

SECT. 3. — *A prudent economy in the spending of their wages by the working classes*, is another matter, essential to their well-being and happiness, to which I would next call your attention.

This consideration, you will perceive, applies to operatives of either sex; and it is a consideration under which, whatever be the rate of wages, a great deal is placed in their power. So much so, that it is a known fact, that with the same wages on which the laborers of one part of England support themselves and bring up their families respectably, those of another part of the country are found to be living in wretchedness. Prices of provision and house rent may, indeed, be higher in the one place than in the other; but this will not explain the difference. For after the advantages and disadvantages of their respective circumstances are fairly balanced, the comparative comfort and respectability of condition will often be found to be essentially different.

As to the wages paid in Bradford, taking the average of every kind of employment, I think it but fair to state my impression, that such average, be it fair or be it low, is generally equal to what is paid in almost any other part of the country. And this impression is derived, not only from inquiries made on the subject from persons within and without the parish, but from the simple and obvious fact, of the floods of laborers who are observed to flow in amongst us whenever there is a demand for extra hands. Of the extent of this influx you may not be fully aware; but I believe that it has been so great, within the last eighteen months or two years, as to have added at least ten thousand to our previously existing large population.

In regard to the rate of wages, I may observe, that it *ought* to be such as to enable the laborer to live, and maintain his family in comfort, and likewise to give him some participation in the advantages of prosperous times.

Yet I need hardly inform you, that the principles which regulate wages, belong more to that class by which are regulated the prices of goods in the market, than of that belonging to moral obligations. Thus labor is a commodity, varying, in its price in the market, according to the demand and supply. If there be more labor in a country than can be duly expended, wages will generally be low; if there be less, as in America, than the demand calls for, wages will of course be comparatively high. Limits, indeed, will be put to the prices of labor by other circumstances. The price, in any country, where labor is too plentiful, will have a check, from the contingent effects, whenever it falls so low that the laborer cannot maintain himself and bring up his family ; and the price, where it is scarce, will be limited from rising extravagantly high, by the check given to the employment of manual labor for purposes that no longer remunerate the employer.

These remarks I am here induced to make, as a caution to our laborers against the fallacies by which designing or speculative persons too often disturb their better feelings, with respect to useless schemes for bettering their condition. Be assured, that no legislative enactment, as some have assumed, could possibly be applied for the regulation of the *amount* of wages. The legislature might as well pretend to fix the price of mercantile commodities. The rate of wages will be the result of the various causes, including also something in regard to human necessities and life, which influence the prices of other things. Wages, in the different departments of labor or handicraft work, will have a fluctuating market price. And that market price, I do firmly believe, cannot be controlled with any permanent advantage, by any unions, or combinations, or turns-out of the laborers themselves. The laborers may, indeed, by such measures, possibly defend themselves against the operation of avarice in individual employers. They may make a stand against arbitrary and unreasonable reductions. But, where the price of labor happens to be a fair market price, determined by the ordinary influence of demand and supply, and the means of reasonable subsistence, measures taken to *force* a rise, will not only certainly fail, but must always result in damage to those the least able to bear a reverse, the laborers themselves.

To my mind, however, it appears that the operative classes — in places especially where 'a fair day's pay is given for a fair day's work' — have a great deal in their power for the bettering of their own condition, and for the advancement of their own independency. And this must be done, by providence and economy, by man and wife, in favorable times. The study of good housewifery in their domestic management, so as to live prudently and carefully in their families, the avoiding of public houses, or useless self-indulgences in food or drink, or gaiety of dress; the laying up of a little store, even by the smallest weekly savings, for the day of adversity, or the time of slack employment — are means more or less within the reach of our laboring population here, by which their condition of independency and comfort may be greatly improved.

What I have elsewhere stated on this point, I would here repeat, that I admire not the position of those operatives who so live from hand to mouth, that, if overtaken by sickness or deprived of work for a short period, they forthwith are reduced to the condition of paupers. Nor do I admire that condition of a working-man which resembles the case of a bankrupt tradesman, or one fit to be a bankrupt, being obliged to sell his stock-in-trade, *his labor*, not as a matter of arrangement with the employer, but, without reference to its value, because he must have money. But this I do admire, and would strongly commend, the prudent adoption of means of self-economy, and mutual help, by which the laborer is defended both from the ordinary chances of destitution and of humiliating degradation.

Thus do I love to see the working man in his proper position *as a man:* not influenced by the spurious species of independency which would incline him to set himself up above those beneath him, or to act disrespectfully towards those above him; but influenced by a just and manly self-consideration of his own equality of nature with the sovereign on the throne, and of his own equality, on Gospel principles and Gospel requirements, in encouragements and hopes for that great and noble destiny in a future state, which an impartial Providence has placed equally before the poor and the rich. This independency of temporal condition, that would enable the laborer to bargain for his labor, or at least to sell it for its market-price value, I hold to be one which would be of vast advantage to the working population, would enable them to occupy their proper position in the scale of being, and which, by introducing a fitting self-respect, would *claim* respect and consideration from others.

SECT. 4. — My next suggestion is for the consideration of the *principals* in our factories. And this I would present to them in the form of the question, — *Whether a more watchful regard to the moral character of their female operatives, and that of those who overlook them, be not practicable?*

Whilst proposing an inquiry of this kind, it is very satisfactory to me to be able, from my own knowledge, to state, that there are mill owners in this town and parish who do interest themselves, with a commendable measure of care, about the moral character and conduct of their operatives. There are those, I believe, who do feel the responsibility of their own position, and who, as masters, will not retain, knowingly, an ill-conducted female in their employ. And there can be no doubt, but that, among the large body of spinners and manufacturers with whose practices I am not acquainted, there are many others who either seek to promote a moral influence, or would be ready, were any feasible plan before them, to aid, by their authority and influence, a measure so benevolent and desirable.

Still, however, it is to be feared that it is the fact, that our master spinners and manufacturers, *as a body*, do by no means feel the responsibility, in regard to the moral character and conduct of their hands, which, on principles of Christian obligation, distinctly attach to them. This observation, which I might have been diffident in putting forth on my personal authority alone, is more extensively supported by the reports of "the commissioners for enquiring into the state of children and young persons employed in our mines, manufactories, and trades," than I have any occasion to assume. In the report on the moral condition of such persons, we find the following, among numerous other remarks to the like effect—that "it is a fearful thing to see, how exempt the employers of labor hold themselves from moral obligations of every description towards those from whose industry their own fortunes spring. Even those who contribute at all to the education or moral improvement of their workmen, do so in nineteen cases out of twenty by money, and without personal pains or superintendence of their own. These vicarious benevolences," adds Mr. Symons, from whose report I give this extract, "are seldom availing." And, "it is pitiable to observe," as some of the witnesses examined state, "what little pains are taken by masters with the morals of the children, and to observe what numbers of young creatures are almost totally neglected." *

But whilst such is represented to be the *general fact* as to mines, manufactures, and trades — for I do not here quote from the factory reports — we have occasional records of a different feeling and consideration on the part of masters, for the moral and religious behavior of those they employ, and, connected with this, we have distinct assurances of a corresponding benefit being produced on the character of their operatives.†

Cases in our own neighborhood, indeed, might be adduced, which fully corroborate this remark. Certain mills in Bradford and in Leeds, a factory at Castlefield, near Bingley, and others, have been named to me, or visited by me, in which the feeling of responsibility on the part of the proprietors, has led to regulations for moral discipline, which, so far as they have been carried, have invariably been followed by the anticipated good results.

But one instance, in another part of the kingdom, has been brought under my notice (just when the manuscript of this section was being corrected for the press) wherein the results of a careful moral discipline were so strikingly satisfactory, and, as such, so eminently adapted for my present object, that I am induced to mention it somewhat fully.

My authority, in the first instance, (verified by a document in the following section,) was a verbal account of a friend, who, in an official capacity, I believe, had occasion to visit and examine into the interesting case.

* "Physical and moral condition of children and young persons employed in mines and manufactures," p. 251.

† Idem, p. 250, &c.

The factory (a silk-factory) is situated at Hadleigh, and was, for some years, under the special care, as to its moral superintendence, of the Archdeacon Lyall. By reason of previous arrangements with the lessee of the works, the condition of a thorough control, as to the moral discipline of the hands, was yielded. This, I understand, went to such an extent, that the committee of moral managers might prevent an ill-conducted person being employed in the mill, and might dismiss from employment any one whose character was found to be unsatisfactory. Discipline, under this peculiar authority, being rigidly exercised by the committee, and the female operatives from the country around being subjected to their control, as to the selection of their lodgings, &c. — the most happy effects on the moral condition, not only of the factory-girls themselves, but, it might be said, of the town at large, ensued. The influence, it was stated, was indeed so salutary and so extensive, that a previously existing indifference to character among many of the lower class of females, gave place to a feeling of shame and reproach on occasions of loose conduct becoming known; whilst the influence was beneficially felt to the extent of the surrounding villages. Hands were, in consequence, obtained more readily and of more respectable classes than the manufacturer had anticipated; generally there were numbers of applicants waiting for their turn; and, ultimately, the daughters of certain tradesmen, who had at first been kept aloof from such employment, became factory operatives. And the overseer, speaking of the good order of the establishment, told my informant, that he could more easily manage the 200 girls, then under his charge, than he could an amount of 60 only at Colchester.

The different degrees of regard which appear to be paid to the moral condition of their operatives, by the manufacturers in this country and those of Lowell, in the United States, I wish by no means to set forth as the fair and equitable comparison of their respective characters, personally, as Christian men. For the manufacturers of Lowell have, besides their proper sense of moral responsibility, influential motives and peculiar conditions, which do not, in like measure, prevail with us.

At the same time, I will be free to say, that we might, with *great advantage* to all parties, emulate their careful moral surveillance. For I know of no feeling more deserving of being cherished, on the part of *masters*, than that of a kind, watchful, and strictly paternal consideration for their work-people — and, particularly, because of their being so exposed to be led astray, for the females in their employ. The caring for their people, not as if they were mere pieces of machinery, but as rational and moral beings, lent to them by the Great Disposer of ranks and conditions among men, for their use, and entrusted to their Christian responsibilities, would greatly conduce to the welfare of both parties, and would serve to cement in kindly and kindred bonds the employers and employed.

SECT. 5. — Another suggestion for the promotion of the object now contemplated, I would propose for the consideration of our manufacturers and others who employ large numbers of females, in the form of this additional inquiry, — *Whether something may not be done for the moral and general benefit of their female operatives*, BEYOND *the walls of the mills and factories?*

Whilst anxiously reflecting on this very interesting and

important subject, it occurred to me, that measures of a very feasible nature might be adopted, calculated, in various ways, for promoting an useful end. And the leading and fundamental measure which I would beg to suggest to our manufacturers, &c. is, — *the formation of an Association among themselves* — uniting with them other benevolently disposed individuals — *for the improvement of the condition of the factory operatives.*

Were such an Association formed, its primary object would be, to consider of the best and most practicable methods for accomplishing the highly important object; and then to organize the fitting instrumentality for carrying the measures agreed upon into effect.

Among the variety of plans which might be devised for an object of this kind, I would suggest, for consideration or deliberation, the following: —

1. The establishment of an Association, as already proposed, of the employers of female labor, and of other friends to the improvement of the moral, intellectual and religious condition of our female operatives.

2. The formation of a general Board, consisting of a selection of manufacturers and spinners, associated with the senior clergy and dissenting ministers of different denominations, and other individuals favorable to the object.

3. The appointment of a visiting committee or committees, comprising a ladies' branch, for visiting from time to time the residences of the female operatives, and especially the lodging-houses, and reporting thereon.

4. The establishment of evening schools, and mutual improvement classes, in different districts of the town, for males as well as females.

5. The establishment of a library of useful books for

the common use, either without payment or on a small subscription, of all the factory operatives employed by the principals connected with the Association.

6. The establishment of a factory-operatives' sick club.

7. The appointment of means for facilitating and encouraging deposits in the Savings' Bank.

This scheme, in its practical details, would embrace the keeping of a register in each mill, or manufacturing establishment, of all their female hands, their ages, conditions (married or single), places of residence and names of the persons, parents or others, with whom they lodge.

The visiting of the residences of the operatives would have for its objects — the determination of the character of the family, and inmates of the house, with whom they respectively reside, — and the promotion, by occasional visits, of cleanliness and good order in the several houses.

The *condition* of *employment* by the principals of factories, &c. would be a *respectable character*, and would involve the approval of the places of abode of the female operatives, or their removing, in the case of lodgings, from residences objected to by the visiters. Dismissal to take place in the event of immoral conduct being ascertained, or habits of immoral conversation being known.

The principle on which the schools and mutual improvement classes should be conducted, must, of course, be such as *to secure general coöperation.* The same as that on which the church and dissenters at present seek to promote education among the children of the poor, is the principle naturally suggested. For, herein, under existing feelings, there can be no union without compromise;

except, indeed, by the total abandonment of a religious character and tone to the instruction conveyed. This system we are not disposed to adopt. My proposal, therefore, is, that, in matters of a mere moral character, all classes and denominations do unite; but, in matters of education or religious instruction, that each part do pursue their own course. We, for our part, will be ready to provide for evening schools and classes on such system as we may deem fitting; let others, who do not coalesce with us, do the same; and let the factory operatives, as free agents, have their unembarrassed choice. We shall thus be secure, I hope, against any dividing influence which might hinder a full and cordial co-operation in the grand enterprise proposed. And, as to myself, I will be free to say, that so acting on clearly understood grounds, I shall truly rejoice in the opportunity of uniting with any and all of my parishioners in a good work which will serve, I trust, at once to benefit our fellow-creatures, and 'to bind the sweet influences' of Christian charity betwixt ourselves.

These are but hints—hints hastily, but not inconsiderately, cast together,—designed to show that there are practical methods, by no means hard to be carried out, in the employment of which, good, under the Divine blessing, must certainly be accomplished. Not, indeed, that we should expect to accomplish all that could be *desired*; but yet, that as reasonable a prospect of benefit would be had, as in the case of the establishment of an hospital or the building of a church. In these cases, all that the most sanguine person could reasonably expect would be, not that every patient brought into the hospital would be cured, nor that every worshipper who might attend the house of God would become a godly person; but that a considerable measure of relief would be afforded to human suffering, or a large measure of good done to the souls of sinners. Though no more, therefore, be assumed to result from the Association proposed, there would be an assurance of general success quite sufficient for the encouragement of the Christian philanthropist.

Of the beneficial effects of such an association on the moral conduct of a population, lamentably, I regret to say, as a body, fallen from the purity which befits, and the virtue which adorns, the female character, there could be no reasonable doubt. A decided system of rebuke to immorality, and of consequent encouragement to chaste and virtuous conduct, could not lack a beneficial result. Worldly interest would be made strongly to bear on incorrectness of conduct. The fear of the loss of employment, on the one hand, with the desire of obtaining employment in the best conducted and most desirable establishments, on the other, would exert a salutary influence. Parental watchfulness would be excited, in cases where now, alas, there prevails too often such deplorable neglect, or indifference, over the moral conduct of daughters, lest the benefit derived from their labors should be lost.

Hence, also, would an additional and influential restraint be placed upon vicious propensities; the ill-conducted female, being detected and dismissed, would be made to lament her error, if not to feel her shame; the tone of morals would thus be raised, and consideration for personal character promoted; and, no doubt, under that blessing from heaven, which on such a righteous cause we should confidently expect, a check would be given to the vice and profligacy so prevalent in this professedly Christian country, whereby vast multitudes of females of the various working classes amongst us have suffered, and do now suffer, such deep degradation, and such great temporal and eternal injury.

In the suggesting of such benefits as these from the exercises of kindly Christian consideration, and systematic attention, on the part of employers and masters, I do not, I believe, by any means overrate the effects on the general character to be reasonably looked for. The case of the Lowell factory girls might justify this remark: but, as a case more parallel with what we have to deal with, that of Hadleigh, already adduced, may be confidently appealed to. If such beneficial effects have been produced elsewhere, why should not effects in some degree the same, be derived from the like moral influence put in operation here?

Such influence, I may moreover observe, though not necessarily *religious*, would act favorably as the pioneer to religion. It would serve to clear the obstacles to their return to the Christian fold of a class of persons, now, as to large numbers, almost entirely estranged. And so might many, now totally out of the reach of the Church, be brought within the influence of a Christian ministry and the institutions of religion, whereby souls would be won to Christ, as well as intelligent rational beings reclaimed from moral and social degradation.

Happy, happy would it be for us and our country, had the gospel, which is so prevalently professed, its legitimate influence amongst our general population. Then would the grand restraining influence, the *fear of God*, become not only effective as a corrective of the outward conduct, but as a purifier of the inward thoughts. Then would the solemn warnings of the bible against licentiousness be felt to be awful truths; the warning, that "whoremongers and adulterers God will judge," would be felt to apply the same to our profligate young men, as to our unchaste, lascivious young women. Then would a population living, alas, for the most part in neglect of their souls, and content with their pitiful portion in this world, be awakened to a perception of their folly, and would be urged to seek, by the salvation in our blessed Lord, and by a godly life and a holy devotedness to Him, for the high destiny and distinction of children of God and heirs of immortal glory!

Thankful should I feel, if the manufacturers and spinners of Bradford should receive these suggestions in the same spirit of kind and Christian earnestness with which I offer them, and might be induced to take up a subject of such honorable import and true benevolence among themselves. And it would be a high honor to Bradford, were it, as a manufacturing town, to come forward in this work of generous consideration for, and kindness towards, the operative instruments of its prosperity, and take that lead in the work of philanthropy, which, by the blessing of God, and the activity of the people, it is taking in enterprise and manufacture! So doing it would consecrate its prosperity, and yield a public and beneficial thank-offering to that gracious Providence, who giveth to men "power to get wealth." [Deut. viii. 18: Hosea ii. 8.]

In closing this practical section,—constituting, in fact, the chief application I desired to make of the Lectures embodied in this little volume,—I have the satisfaction of still further supporting my own impressions of the feasibleness of the suggested plan by an interesting communication, received almost at the very moment when required for the press, from Archdeacon Lyall, to whom I had written for the verification or correction of the sketch of the Hadleigh factory given in Sec. 4.

The letter is so important to my object, that, hastily as the communication was obliged to be made, I will venture (under the impression that I shall be sanctioned in so doing) to introduce it in this place.

"Canterbury, March 24th, 1845.

"Dear Sir,

"Your letter was forwarded to me at this place, and only reached my hands yesterday before service. I am very sorry that it came to me at a moment when I am much engaged, requiring, as it does, so immediate an answer.

"The history of the Hadleigh Silk Factory is very short and simple. It was built in 1834 by the Parishioners, in shares of £50, and let on lease to a large silk manufacturer, Mr. Stephen Brown, of Colchester, for 21 years at 6 per cent. It was to employ 400 children. The trustees being the rector and churchwardens, a power was reserved to them, and such persons as they might name, to the following effect:—That no children should be admitted except by their appointment:—That they should also have power to suspend or dismiss any who should misconduct themselves either in or out of the factory. Also a certain weekly sum was retained out of their wages, for the payment of a school-mistress. Any superintendent or mechanic employed about the premises, whether male or female, behaving immorally, was liable to be removed by the committee; but the appointment of these last was in the hands of the proprietor.

"The Committee met once or twice, but after that they left the management entirely in my hands: so that, in fact, the above powers devolved altogether upon the rector and his curate only.

"Under these powers, a variety of minor regulations were agreed upon.—Any young woman having a child, was dismissed, and never re-admitted, under any circumstances. Neither was any young person ever admitted who had ever had a child. A report was brought to me every morning, containing the names of any who had been seen walking in the streets at improper hours, or in improper company, or found in any houses of public entertainment. Those from the country were allowed only to lodge in houses approved by me, under pain of being suspended from their work or dismissed.

"I found not the least difficulty or trouble in carrying these regulations out, on the part of the children, though I had a good deal at first, from some other quarters. After a few months, however, all worked smoothly, and the care of the factory was no trouble at all, but rather the reverse. For the last two years of my residence at Hadleigh, which I quitted in 1842, I had no occasion to dismiss a single girl, and very rarely any occasion even to admonish, except for mere childish offences.

"Of its beneficial influence upon the parish in a moral point of view, there could be no question; and when I state that the rates, before the factory, were fifteen shillings in the pound, and that when I left they were not more than from three to four shillings in the pound, it is unnecessary to say more of the benefits derived by the rate-payers.

"How well it must have answered to the owner of the mill, is proved by the fact, that after a few months' experience of its working, he called upon me to ask, if I would use my influence with the clergy at Colchester, to assume the same powers in his larger factory there, which I had exercised at Hadleigh. And a still stronger proof was afterwards afforded:—Being desirous of building a factory in the parish of Nayland, about eight miles off, he told me that if I would persuade the clergymen there, to superintend the mill, as I had done, he would make the land on which it was to be built over to me, and take it from me at a nominal rent upon a lease of 99 years, *inserting in the lease all the same provisions as were introduced into the lease of the Hadleigh-Mill.* The Mill was afterwards built, but, for reasons not necessary here to be stated, without these clauses.

"It is my opinion that every one of the regulations introduced by me at Hadleigh might with ease be acted upon in every factory, whether in a town or in the country. And provided they gave me no more trouble than mine did at Hadleigh after the first year and a half, I would undertake, with the assistance of the owners, to superintend all the Mills in Leeds or Bradford, or any town in England. I am persuaded that if any mill-owner was to make the experiment, his example would soon be followed, as it would be found that his mill would always have the children of the most respectable parents in the place, in preference to any other, and be managed both cheaper and better, in the same proportion.

"Mr. Saunders's report, which you have sent to me,

contains, I think, a very true account of my Hadleigh mill, and of the benefits it conferred both on the children and the town and neighborhood.

"Believe me to be, my dear Sir,

"Very truly yours,

"W. R. LYALL."

SECT. 6.—One more requirement, as I believe it to be, for the placing of our operatives, *generally*, in their proper position, is,—*The abridging of the hours of labor.*

This being a subject on which the views of the employers and the employed, as distinct bodies, must be considered, in certain respects, at issue,—it could not be made a part of any plan, for general adoption, for the object I have in view. To have introduced it, therefore, as an element in the foregoing plan, would have been to destroy its practicability, and to cause it to be postponed to such time — however indefinite the period — when the boon so much longed for by our working population, and many of our Christian philanthropists, should have been attained.

As it is a subject, however, to which the public mind has been so strongly turned by reason of the agitation of the question in parliament, as well as amongst us, and elsewhere, in the country, that I could not satisfy myself without stating my own views thereon, on an occasion like this, when I have so many of our working population, besides employers, before me.

The consideration, let it be observed, of the long hours of labor, is one by no means belonging peculiarly to our factory-system. The Reports of the Commissioners appointed by the government, 'for inquiring into the employment of children and young persons in mines and collieries, trades and manufactures,' have brought to light a most appalling amount of misery induced, as to this one essential element, by over-working. And from hence we find, taking the whole range of the investigations under this humane commission, that the oppression of the laborer is no local or peculiar incident, but an evil of huge magnitude, as a national sin. It is an evil which has grown up insidiously amongst us, the offspring of success in trade and of an excess of laboring population.

National prosperity has stirred up a mighty spirit of commercial enterprise. The advance in wealth has stimulated competition. The growth of luxury and ostentatious living, springing out of singular prosperity in the land, have generated a reactive desire, amounting to the feeling of necessity, for the acquisition of wealth. Thus enterprise has been pushed to the utmost. The pursuance of riches has become all-absorbing. But, we pause to ask, by what means, and at what cost, are these results obtained? Too often, alas, as the obvious fact appears, by sacrificing to Mammon, instead of to God, on the one part, and by the enslavement to almost uninterrupted toil, in regard to large masses of the laboring population, on the other part.

These remarks, you will remember, apply generally, to what must be considered as a great and national evil,—the forcing of national aggrandisement at the expense of the health and comfort, too often of the life and souls, of the over-worked laborers! Yet, whilst we make these remarks, we must be painfully conscious, that, as a general fact, the evil is far more manifest than the remedy. In many cases no effective remedy can be applied by the legislature, and no relief, it is feared, can be yielded but by means of individual consideration and Christian kindness.

In the special case of factories, however, with which we have more particularly to deal, where such numbers of children and young persons are employed, the remedy, so far as the abridging of the hours of labor goes, has, to the extent of the existing regulations, been already tried. But beyond what has been done, it is my clear impression, in agreement with the views of multitudes of my countrymen, that more needs to be done, and that more might be done.

In this particular, Lowell, which has guided us in several of the foregoing suggestions, affords us no example, as their hours of factory labor are fully as long as ours. But if their factory-operatives are found in a better and more respectable position, than the like class in England, it is not because of their long hours, but in spite of them. The other causes, on which I have remarked, are sufficient to account for all the differences. And, I may add, if they are now well off with long hours, they would, physically and socially, if not mentally and morally, be better off with reduced and more reasonable hours. And that which I consider an error in the factory system of Lowell, I believe to be a much greater error with us. For the factory operatives with us begin younger, and do not start fair with them in point of education. We have more actual need, therefore, than they have, of a long evening. We have more to get up, more to acquire.

More married women, too, work in factories with us than at Lowell. I wish from my heart there were fewer. No woman, with a young family, can possibly do justice to them whilst working abroad. Nay, no married woman can make her husband really comfortable and keep his house respectable, whilst working herself as a factory operative. For the clean and orderly cottage, the cheerful fire-side, the pleasant, wholesome fare, and the smiling, happy wife, there must be a stay-at-home wife. And better, far better, I would say, is the humblest fare, if it be but wholesome and sufficient, with so many elements ministering to domestic comfort, than the gain of a wife's toil (except necessity require) throughout the day at the mill, with hardly a temporal circumstance left to make a hard-working man happy!

As to the shortening of the hours of labor, the time, I think, is not far distant, when this will be effected. Manufacturers themselves, indeed, we can hardly expect as a body to agree voluntarily to limit the time of their working. And individual manufacturers cannot work at shorter time than others, and pay the same wages, but at an obvious disadvantage. The fact is, therefore, demonstrable, that this boon cannot be obtained for our working population but by legislative interference. For adults, indeed, no law can apply. The legislature cannot interfere with any man's liberty to work as long as he pleases. But in the case of children and young persons, and females, where such are not in the position of free-agents, the legislature may justly, as it already benevolently has done, step in with a paternal consideration, and official authority, for their protection. But, practically, so far as many of our factories are concerned, it is well understood that the limiting of the time of young persons, would tend to regulate and limit that of the adults.

As to the *specific period* for which it is fitting and proper that the laborer should pursue his toils, there has been, and of course still will be, considerable difference of opinion. When I first came among you, to superintend the spiritual concerns of the church in this parish, I am free to say, that I could not quite see how the cardinal number of "ten hours" was discovered and verified. But in the progress of my inquiries on this important question, my mind has been gradually cleared, until I have arrived, as my personal conviction, at this conclusion, *that twelve hours a day for labor, including intervals for meals, is the fitting and natural limits of a day's labor.* The average day-light is twelve hours. We are reminded in scripture that 'there are twelve hours in the day.' And the scriptural intimation of *duty* is 'to work while it is day: for the night cometh when no man can work,' or (might we not so appropriate the saying?) '*ought* to work.'

It being an admitted fact, too, that our general population is over-worked, as, also, that an over-worked population is short-lived, a limitation of the average period of labor is clearly called for. Taking, moreover, the average strength of the human constitution, I apprehend that the period of labor of twelve hours (inclusive of meals) will be found as much as can be devoted, day by day, to any one business, or description of toil, without injury to the constitution, and the abridgment of life of the laborer.

Some object to a law applying to factories, and not reaching the general case of an over-worked population. To this partiality I should also object, if the restriction of the case referred to, would, on the general balance of advantages and disadvantages, be certainly injurious. But my decided conviction is, that it would not. For where our labor is in excess, and our production in excess too, should we not have a more healthy condition of trade? In such case, if profits were not so large, would they not be more steady? If the full tide of prosperous times should not rise so high, would it not happen, according to the general law of nature, that the ebb tide of depression would not fall so low? So that, on the whole, might we not expect, that neither manufacturers in their profits, nor operatives in their wages, would suffer material damage?

Other objections are wont to be urged against a further limitation of labor, for young persons and women in our factories, than that which at present exists, the limitation to a clear period of twelve hours, actual work, which, had the subject now touched upon been the essential one of these lectures, it would have been needful to have considered. Among the objections to a reduction of working-hours, we not unfrequently hear such as the following: — that the measure would give more idle time for the laborers to be spent in dissipation or vice; that it would necessarily reduce their wages in the same proportion as the time was reduced; that it would incapacitate the manufacturer, in times of great demand for his fabrics, from fulfilling his orders, and so throw a proportion of the trade into foreign hands: and that, by increasing the price of production, it would yield up the advantage, which, as a manufacturing country, Britain now possesses, in favor of foreign competition.

Though these several objections, besides others, might, on sound principles of political economy, be shown, I think, to be of far inferior importance to what is assumed; yet I may here more satisfactorily and more simply meet them by taking a higher ground for their consideration.

If our laboring population be over-worked, the Christian economist and philosopher should feel no anxiety and apprehend no risk, in doing towards them, and for them, that which is right. If it be the blessing of the Lord that giveth prosperity to men and nations, as by the sacred authorities already adduced has been shown, then we might reasonably calculate upon a *greater blessing* in doing what is right, than in pursuing what is wrong. No one can fairly and rationally look upon the growing enterprise and extension of manufactures in America, and on the continent of Europe, without some serious apprehension of a decline in our present superiority. If, on principles of worldly policy, then, the prospect before us be less encouraging than what it has hitherto been; surely our greatest safeguard would be in the laying of ourselves out for the Divine blessing, by taking the high ground of kindness and humanity, and of Christian consideration for our fellow creatures!

Advocating, however, as I earnestly would, an abridgment of the long hours of labor in our factories, I am far from approving of the spirit in which this question has too often been dealt with. On the contrary, I can safely take you to witness, whether in my repeated attendance at your meetings for promoting this object, I have not only deprecated such a spirit, but have ever striven to set before you principles, calculated for the promotion of mutual consideration and kindly feeling betwixt the operatives and their employers. And I will be free to assert, and to maintain the assertion on all fitting occasions, that those men are no friends to the working classes, but, practically, their greatest enemies, who in schemes for promoting their advantage assume the existence of opposition, or diversity of interests, betwixt men and masters. Their true interests are identical; and true kindness and true patriotism alike claim of both classes the cultivation of mutual confidence and reciprocal good will.

There are classes of our population, engaged in different objects of pursuit, who are too apt to view each other with ungenerous and suspicious feelings. Thus manufacturers and agriculturalists are, unfortunately, wont to show, as classes, little tenderness for each other. But whatever be the faults of either, or of those other classes who so loudly censure them, of this I am persuaded, that, let the censurers change places, and, under the same circumstances, they would pursue like courses. Human nature is essentially similar, and essentially selfish. And the desire of gain, where pursued under mere natural influences, will ever run into the same errors and excesses, no matter who, amongst mere worldly men, is the adventurer! True self-denying religion can alone chasten the inordinate love of gain; and till that sacred influence be prevalent and penetrating in the world of business, the business-world will continue to be characterized by selfishness and the love of Mammon!

APPENDIX.

The foregoing lectures are republished in this country because they are written with a remarkable fidelity of description, and are free from those errors which so generally mark the writings of travellers. They have been read with interest in Great Britain, and we presume, from the fact that the Hon. Mr. Everett accompanied a copy of them, which he sent to a friend here, with the remark, "that it was the only copy he could find," that the English edition was eagerly sought after.

Were it in our power we should give a sketch of the life and labors of Dr. Scoresby in the appendix to the American Edition, since his history and his services, if they were known, would add to the interest of his writings. It may not be superfluous to say that he was thoroughly educated by his father, Captain Scoresby, for the sea, and at an early age commanded a ship on a voyage to the Northern ocean. He afterwards built a ship at his own expense, and having made every preparation for a voyage of whaling and discovery, and manned her with a crew of fifty men, he set sail for the coast of Greenland. He penetrated into those dangerous seas, and though steadily and successfully engaged in whaling, he pushed his scientific researches with great vigor. The result he gave to the public on his return home in an interesting volume dedicated to the King. He afterwards turned his attention from the sea, and pursued various studies to which his taste inclined him. But especially he became interested in the study of Theology, to which he finally devoted himself as the means by which he could do the most good. Having an independent property, he gave his whole time to the pursuit of truth, and to the amelioration and improvement of the condition of his countrymen. For several years he has held the vicarage of Bradford in Yorkshire, which includes the charge of 60,000 souls, and he has spared no effort to introduce and carry out many important plans for their moral and intellectual improvement. The lectures, of which the foregoing are a part, were delivered before an association of his parishioners, and were not intended for publication. But this portion of them has been published at the request of some owners of Manufacturing property in the district, with the design of exciting, if possible, some desire both among the operatives and principals to bring about a better state of things. We regret to learn that Dr. Scoresby has signified his intention of resigning his present charge ; finding it, as he says, too much labor and responsibility for him to undergo without the loss of his health.

In his visit to this country he was indefatigable in his efforts to obtain correct knowledge on all subjects. His opportunities for acquiring information were unusually good, since his striking simplicity of character, his easy and unassuming manners, and his agreeable and learned conversation made him friends whereever he went. He was very minute in his collection of facts ; his early habits and studies made him almost necessarily correct, particularly in figures ; it is remarkable that his lectures, which were all puplished in the English papers as they were delivered, contain few or no mistakes even in names.

The increase of mills in Lowell since the Author's visit last year is unusually large. One new Company has been organized with a capital of $600,000, a large mill has been added to the works of one Company, additions have been made to two others, and two large new mills are to be erected by two more. The waters of the Merrimack are to be used again below Lowell, where a fall of from 23 to 25 feet is to be obtained by the erection of the dam of the Essex Co., which is going into operation with a capital of $1,000,000. Here a new town is to be laid out, which at some future time may nearly equal Lowell in size.

The ratio of increase in other manufacturing towns is equally great. At Amoskeag, Saco, Nashua, Salmon Falls, Newburyport, Salem, Fall River, and other smaller places in New England, and in many places in other States, new mills are erecting. Several new kinds of manufacture also have been introduced with every promise of ultimate success.

That the present prosperity can continue, it is unreasonable to expect, since the accumulation of goods must soon bring with it a corresponding depression ; but whatever revulsion may follow, the present impulse will at least have the effect of placing the manufacturing interest of the country on a basis from which it cannot be easily moved by hostile legislation. Neither is there any reason to believe, if the system of taking care of the moral condition of the operatives should be introduced where it does not exist already, and should be continued where it does, that the character of our people will degenerate, or their true happiness be diminished, while the wealth of our country is increased.

A. A. L.

The following Statistics of Lowell Manufactures are added,

Statistics of Lowell Manufactures, January 1,

CORPORATIONS.	Locks & Canals.	Merrimack.	Hamilton.	Appleton	Lowell.
Incorporated,	1792	1822	1825	1828	1828
Commenced operations,	1822	1823	1825	1828	1828
Capital Stock,	600,000	2,000,000	1,000,000	600,000	600,000
Number of Mills,	2 Shops, Smithy, Furnace.	5 & PrintWks	3 & PrintWks	2	2, 1 cotton, 1 carpet.
Spindles,		40,384	21,248	11,776	6,000 Cotton, beside Wool.
Looms,		1,300	590	400	152 Cotton, 50 power carpet, 40 hand do.
Females employed,		1,250	650	340	400
Males employed,	500	550	250	65	200
Yards made per week,	1225 tons wr't and cast iron per annum.	250,000	100,000	100,000	2,500 Car. 150 Rugs, 85,000.
Bls. of Cotton used in do.,		130	100	90	110
Lbs. of Cotton wro't in do.	Machinery, R R Cars & Engines.	56,000	42,000	36,000	40,000
Yds. dyed and printed do.,		210,000	63,000		
Kind of Goods made,	15,000 bush. charcoal, 200 chal. smith's coal, 400 tons hard coal.	Pr'ts & Sheetings, No. 22 to 40.	Prts, Flannels Sheetings &c No. 14 to 40.	Sheet. & Shirt. 14.	Carpets, Rugs & Negro Cl'th
Tns. Anthracite Cl. pr an.		5,000	3,000	300	500
Cords of Wood per ann.,	200	200	500		500
Galls. of Oil per annum,	2,300	13,000	6,500	3,440	Olive, 4,000. Sperm, 4000.
Diameter of Water-whls.,	13 ft.	30 ft.	13 ft.	13 ft.	13 ft.
Lgth of do. for each mill,	14 ft.	24 ft.	42 ft.	42 ft.	60 ft.
How Warmed,	Hot Air Furn.	Steam.	Steam & H.A.	Steam.	Hot Air Furn.

Yards of Cloth per annum, ... 74,141,600
Pounds of Cotton consumed, ... 22,880,000
Assuming half to be Upland, and half New Orleans and Alabama, the consumption in bales, 361 lbs. each, is . . . . , 58,240
A pound of Cotton averages 3 1-5 yards.
100 lbs. Cotton will produce 89 lbs. Cloth.
Average wages of Females, clear of board, per week, . . . $1.75
Average wages of Males, clear of board, per day, ... 70
Medium produce of a Loom, No. 14 yarn, yards per day, . . . 44 to 45
" " " No. 39 " " " 30
Average per Spindle, yards per day, ... 1 1-10
Average amount of wages paid per month, ... $150,000
Consumption of Starch per annum, (lbs.) 800,000
Consumption of Flour for Starch in Mills, Print Works, and Bleachery, bbls. per annum, ... 4,000
Consumption of Charcoal, bushels per annum, ... 600,000

which are not contained in the original volume.

1844. Compiled from Authentic Sources.

Middlesex.	Suffolk.	Tremont.	Lawrence.	Boott.	Mass'ts.	Total.
1830	1830	1830	1830	1835	1839	
1830	1832	1832	1833–4	1836	1840	
750,000	600,000	600,000	1,500,000	1,200,000	1,200,000	10,650,000
2 and 2 Dye-houses.	2	2	5	4	4	33, exclusive pr't-wks, &c.
7,200	11,776	11,520	32,640	31,524	27,008	201,076
37 Br'dcloth, 122 Cassim.	352	409	950	910	882	6,194
550	340	360	900	780	725	6,295
250	70	70	170	130	160	2,345
9,000 Cassim. 1,800 Br'dcl.	90,000	115,000	210,000	180,000	260,000	1,425,800
1,000,000 lbs. wool per ann.	90	75	180	145	200	1,120
and 3,000,000 teasels.	32,000	30,000	65,000	59,000	80,000	440,000
Broadcloth & Cassimere.	Drill. 14.	Sheet. & Shirt. 14.	Print'g Cl'ths Sheet & Shirt No. 14 to 30.	Drill. 14, Shirt. 40. Pr'gt'140.	Sheet 13. Shirt. 14. Drill. 14.	273,000
600	300	250	650	750	750	12,500
1,500	70	60	120	70	70	3,290
Lard, 15,000. Sperm, 5,000.	3,500	3,692	8,217	7,100	7,100	67,849
17 & 21 ft.	13 ft.	13 ft.	17 ft.	17 ft.	17 ft.	
23 & 21 ft.	42 ft.	42 ft.	60 ft.	60 ft.	60 ft.	
Fur.& Steam.	Steam.	Steam	Steam.	St & H.A	Steam.	

The Locks and Canals Machine Shop, included among the 33 Mills, can furnish Machinery complete for a Mill of 5000 Spindles in four months; and lumber and materials are always at command, with which to build or rebuild a Mill in that time, if required. When building Mills, the Locks & Canals Company employ directly and indirectly from 1000 to 1200 hands.

To the abovenamed principal establishments may be added, the Lowell Water-Proofing, connected with the Middlesex Manufacturing Company; the extensive Powder Mills of O. M. Whipple, Esq.; the Lowell Bleachery, with a capital of $50,000; Flannel Mill; Blanket Mill; Batting Mill; Paper Mill; Card and Whip Factory; Planing Machine; Reed Machine; Foundry; Grist and Saw Mills — together employing about 500 hands, and a capital of $500,000.

With regard to the health of persons employed in the mills, six of the females out of ten enjoy better health than before entering the mills; and of the males one-half derive the same advantage. In their moral condition and character they are not inferior to any portion of the community.

DIVIDENDS OF MANUFACTURING COMPANIES.

Dividends of Lowell Manufacturing Companies.

	Merrimack.	Hamilton.	Appleton.	Lowell.	Suffolk.	Middlesex.*	Tremont.	Lawrence.	Boott.	Massachusetts.
1839,	11	†	5	5‡	11		11	10	11	
1840,	9	5	5	none.	8		7	3	4	
1841,	12	8	6	11	11	10	8	7	11	‖
1842,	8	none.	none.	4§	3	9	2	2	3	3
1843,	16	6	6		6	4¶	6	7	5	4
1844,	10	7	6	7	16	10	16	15	10	14

* Not in operation as a distinct corporation till 1841.

† Dividend in new shares.

‡ And dividend in new shares.

‖ Not in full operation.

§ The cotton mill was stopped from July 1842 to July 1843.

¶ With an extra dividend of accumulated profits on manufacturing sales of land, and rents, of 25 per cent., to increase the capital from $600,000 to $750,000.

Other Manufactories in New England.

	Cabot.	Chicopee.	Dwight.	Perkins.	Thorndike.	Palmer.	Otis.†	Amoskeag.	York.	Nashua.
1839,	6	9	*	5		20			16	10
1840,	3		*	5		8			12	none.
1841,	9	3	*	6	11	10		6	9	8
1842,	5		3		3	6		4	7	3
1843,	11		11	9	5	9	10	7	6	6
1844,	10	3	8	10	14	16	10	9	17‡	8

* Not in operation.

† The earnings of this Company from 1839 to 1843, were expended in adding to, and renovating the machinery and properties of the Corporation.

‡ And an extra dividend of 10 per cent. from the reserved profits of former years.